Succeed

Eureka Math®
Grade 5
Modules 1 & 2

Published by Great Minds®.

Copyright © 2018 Great Minds®.

Printed in the U.S.A.

This book may be purchased from the publisher at eureka-math.org.

10 9 8 7 6 5 4 3

ISBN 978-1-64054-093-4

G5-M1-M2-S-06.2018

Learn ◆ Practice ◆ Succeed

Eureka Math® student materials for *A Story of Units*® (K–5) are available in the *Learn, Practice, Succeed* trio. This series supports differentiation and remediation while keeping student materials organized and accessible. Educators will find that the *Learn, Practice,* and *Succeed* series also offers coherent—and therefore, more effective—resources for Response to Intervention (RTI), extra practice, and summer learning.

Learn

Eureka Math Learn serves as a student's in-class companion where they show their thinking, share what they know, and watch their knowledge build every day. *Learn* assembles the daily classwork—Application Problems, Exit Tickets, Problem Sets, templates—in an easily stored and navigated volume.

Practice

Each *Eureka Math* lesson begins with a series of energetic, joyous fluency activities, including those found in *Eureka Math Practice.* Students who are fluent in their math facts can master more material more deeply. With *Practice,* students build competence in newly acquired skills and reinforce previous learning in preparation for the next lesson.

Together, *Learn* and *Practice* provide all the print materials students will use for their core math instruction.

Succeed

Eureka Math Succeed enables students to work individually toward mastery. These additional problem sets align lesson by lesson with classroom instruction, making them ideal for use as homework or extra practice. Each problem set is accompanied by a Homework Helper, a set of worked examples that illustrate how to solve similar problems.

Teachers and tutors can use *Succeed* books from prior grade levels as curriculum-consistent tools for filling gaps in foundational knowledge. Students will thrive and progress more quickly as familiar models facilitate connections to their current grade-level content.

Students, families, and educators:

Thank you for being part of the *Eureka Math®* community, where we celebrate the joy, wonder, and thrill of mathematics.

Nothing beats the satisfaction of success—the more competent students become, the greater their motivation and engagement. The *Eureka Math Succeed* book provides the guidance and extra practice students need to shore up foundational knowledge and build mastery with new material.

What is in the Succeed *book?*

Eureka Math Succeed books deliver supported practice sets that parallel the lessons of *A Story of Units®*. Each *Succeed* lesson begins with a set of worked examples, called *Homework Helpers*, that illustrate the modeling and reasoning the curriculum uses to build understanding. Next, students receive scaffolded practice through a series of problems carefully sequenced to begin from a place of confidence and add incremental complexity.

How should Succeed *be used?*

The collection of *Succeed* books can be used as differentiated instruction, practice, homework, or intervention. When coupled with *Affirm®*, *Eureka Math*'s digital assessment system, *Succeed* lessons enable educators to give targeted practice and to assess student progress. *Succeed*'s perfect alignment with the mathematical models and language used across *A Story of Units* ensures that students feel the connections and relevance to their daily instruction, whether they are working on foundational skills or getting extra practice on the current topic.

Where can I learn more about Eureka Math *resources?*

The Great Minds® team is committed to supporting students, families, and educators with an ever-growing library of resources, available at eureka-math.org. The website also offers inspiring stories of success in the *Eureka Math* community. Share your insights and accomplishments with fellow users by becoming a *Eureka Math* Champion.

Best wishes for a year filled with Eureka moments!

Jill Diniz

Jill Diniz
Director of Mathematics
Great Minds

Contents

Module 1: Place Value and Decimal Fractions

Module 2: Multi-Digit Whole Number and Decimal Fraction Operations

Topic G: Partial Quotients and Multi-Digit Decimal Division

Topic H: Measurement Word Problems with Multi-Digit Division

Grade 5
Module 1

Note: It is common to encourage students to simply "move the decimal point" a number of places when multiplying or dividing by powers of 10. Instead, encourage students to understand that the decimal point lives between the ones place and the tenths place. The decimal point does not move. Rather, the digits shift along the place value chart when multiplying and dividing by powers of ten.

Use the place value chart and arrows to show how the value of the each digit changes.

1. $4.215 \times 10 =$ **42.15**

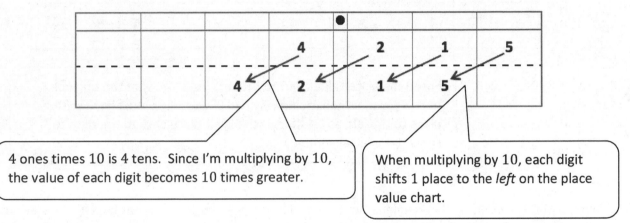

4 ones times 10 is 4 tens. Since I'm multiplying by 10, the value of each digit becomes 10 times greater.

When multiplying by 10, each digit shifts 1 place to the *left* on the place value chart.

2. $421 \div 100 =$ **4.21**

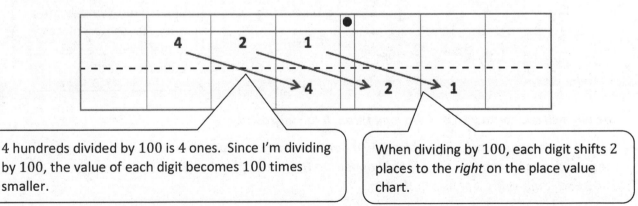

4 hundreds divided by 100 is 4 ones. Since I'm dividing by 100, the value of each digit becomes 100 times smaller.

When dividing by 100, each digit shifts 2 places to the *right* on the place value chart.

EUREKA
MATH

Lesson 1:   Reason concretely and pictorially using place value understanding to
             relate adjacent base ten units from millions to thousandths.

© 2018 Great Minds®. eureka-math.org   3

3. A student used his place value chart to show a number. After the teacher instructed him to multiply his number by 10, the chart showed 3,200.4. Draw a picture of what the place value chart looked like at first.

> 3 hundreds times 10 is 3 thousands. The original number must have had a 3 in the hundreds place.

thousands	hundreds	tens	ones	•	tenths	hundredths	thousandths
3	2	0	.		0	4	

> I used the place value chart to help me visualize what the original number was. When multiplying by 10, each digit must have shifted 1 place to the left, so I shifted each digit 1 place back to the right to show the original number.

4. A microscope has a setting that magnifies an object so that it appears 100 times as large when viewed through the eyepiece. If a small bug is 0.183 cm long, how long will the insect appear in centimeters through the microscope? Explain how you know.

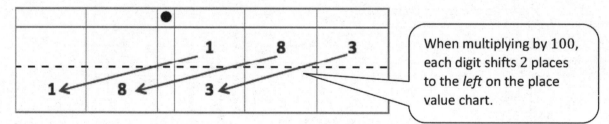

> When multiplying by 100, each digit shifts 2 places to the *left* on the place value chart.

The bug will appear to be **18. 3 cm** *long through the microscope.*

Since the microscope magnifies objects **100** *times, the bug will appear to be* **100** *times larger. I used a place value chart to show what happens to the value of each digit when it is multiplied by* **100.** *Each digit shifts* **2** *places to the left.*

Lesson 1:   Reason concretely and pictorially using place value understanding to relate adjacent base ten units from millions to thousandths.

© 2018 Great Minds®. eureka-math.org

Name _____   Date _____

1. Use the place value chart and arrows to show how the value of each digit changes. The first one has been done for you.

    a.   4.582 × 10 = ___45.82___

    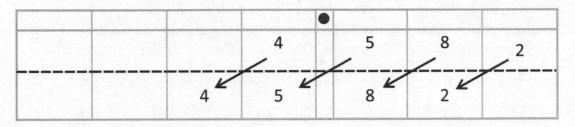

    b.   7.281 × 100 = _____

    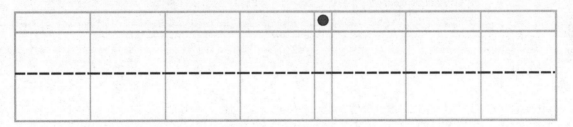

    c.   9.254 × 1,000 = _____

    d.   Explain how and why the value of the 2 changed in (a), (b), and (c).

Lesson 1:   Reason concretely and pictorially using place value understanding to relate adjacent base ten units from millions to thousandths.

© 2018 Great Minds®. eureka-math.org

5

2. Use the place value chart and arrows to show how the value of each digit changes. The first one has been done for you.

   a. 2.46 ÷ 10 = _____0.246_____

   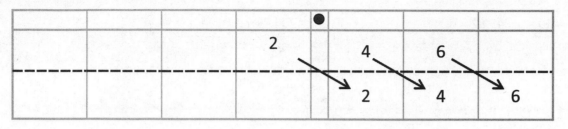

   b. 678 ÷ 100 = _____

   c. 67 ÷ 1,000 = _____

   d. Explain how and why the value of the 6 changed in the quotients in (a), (b), and (c).

Lesson 1:   Reason concretely and pictorially using place value understanding to relate adjacent base ten units from millions to thousandths.

© 2018 Great Minds®. eureka-math.org

3. Researchers counted 8,912 monarch butterflies on one branch of a tree at a site in Mexico. They estimated that the total number of butterflies at the site was 1,000 times as large. About how many butterflies were at the site in all? Explain your thinking, and include a statement of the solution.

4. A student used his place value chart to show a number. After the teacher instructed him to divide his number by 100, the chart showed 28.003. Draw a picture of what the place value chart looked like at first.

				•			

Explain how you decided what to draw on your place value chart. Be sure to include reasoning about how the value of each digit was affected by the division.

5. On a map, the perimeter of a park is 0.251 meters. The actual perimeter of the park is 1,000 times as large. What is the actual perimeter of the park? Explain how you know using a place value chart.

EUREKA MATH

Lesson 1:   Reason concretely and pictorially using place value understanding to relate adjacent base ten units from millions to thousandths.

© 2018 Great Minds®. eureka-math.org

7

1. Solve.

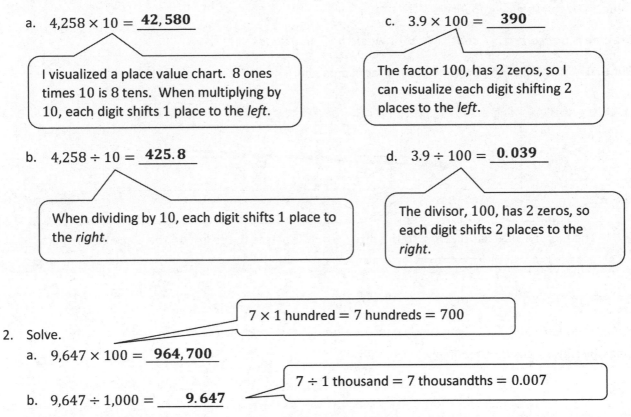

a. $4,258 \times 10 = \underline{\textbf{42,580}}$

I visualized a place value chart. 8 ones times 10 is 8 tens. When multiplying by 10, each digit shifts 1 place to the *left*.

c. $3.9 \times 100 = \underline{\textbf{390}}$

The factor 100, has 2 zeros, so I can visualize each digit shifting 2 places to the *left*.

b. $4,258 \div 10 = \underline{\textbf{425.8}}$

When dividing by 10, each digit shifts 1 place to the *right*.

d. $3.9 \div 100 = \underline{\textbf{0.039}}$

The divisor, 100, has 2 zeros, so each digit shifts 2 places to the *right*.

2. Solve.

$7 \times 1 \text{ hundred} = 7 \text{ hundreds} = 700$

a. $9,647 \times 100 = \underline{\textbf{964,700}}$

$7 \div 1 \text{ thousand} = 7 \text{ thousandths} = 0.007$

b. $9,647 \div 1,000 = \underline{\textbf{9.647}}$

c. Explain how you decided on the number of zeros in the product for part (a).

I visualized a place value chart. Multiplying by 100 *shifts each digit in the factor* $9,647$ *two places to the left, so there were* 2 *additional zeros in the product.*

d. Explain how you decided where to place the decimal in the quotient for part (b).

The divisor, $1,000$*, has* 3 *zeros, so each digit in* $9,647$ *shifts* 3 *places to the right. When the digit* 9 *shifts* 3 *places to the right, it moves to the ones places, so I knew the decimal point needed to go between the ones place and the tenths place. I put the decimal between the* 9 *and the* 6*.*

 EUREKA MATH®   Lesson 2:   Reason abstractly using place value understanding to relate adjacent base ten units from millions to thousandths.   9

© 2018 Great Minds®. eureka-math.org

3. Jasmine says that 7 hundredths multiplied by 1,000 equals 7 thousands. Is she correct? Use a place value chart to explain your answer.

   Jasmine is not correct. 7 ones × 1,000 would be 7 thousands.

   But 0.07 × 1,000 = 70. Look at my place value chart.

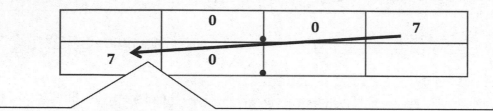

   The factor 1,000 has 3 zeros, so the digit 7 shifts 3 places to the left on the place value chart.

4. Nino's class earned $750 selling candy bars for a fundraiser. $\frac{1}{10}$ of all the money collected was from sales made by Nino. How much money did Nino raise?

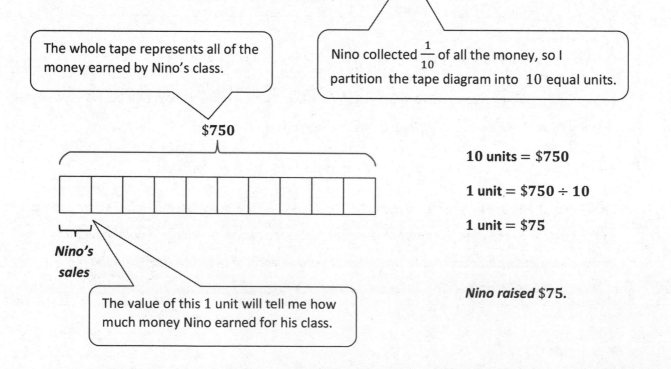

   The whole tape represents all of the money earned by Nino's class.

   Nino collected $\frac{1}{10}$ of all the money, so I partition the tape diagram into 10 equal units.

   $750

   Nino's sales

   The value of this 1 unit will tell me how much money Nino earned for his class.

   10 units = $750

   1 unit = $750 ÷ 10

   1 unit = $75

   Nino raised $75.

Lesson 2:   Reason abstractly using place value understanding to relate adjacent
                                          base ten units from millions to thousandths.

Name _____   Date _____

1. Solve.

    a.   36,000 × 10 = _____   e.   2.4 × 100 = _____

    b.   36,000 ÷ 10 = _____   f.   24 ÷ 1,000 = _____

    c.   4.3 × 10 = _____   g.   4.54 × 1,000 = _____

    d.   4.3 ÷ 10 = _____   h.   3,045.4 ÷ 100 = _____

2. Find the products.

    a.   14,560 × 10   = _____

    b.   14,560 × 100   = _____

    c.   14,560 × 1,000 = _____

    Explain how you decided on the number of zeros in the products for (a), (b), and (c).

Lesson 2:   Reason abstractly using place value understanding to relate adjacent
             base ten units from millions to thousandths.

© 2018 Great Minds®. eureka-math.org

11

3. Find the quotients.

    a. 16.5 ÷ 10   = _____

    b. 16.5 ÷ 100 = _____

    c. Explain how you decided where to place the decimal in the quotients for (a) and (b).

4. Ted says that 3 tenths multiplied by 100 equals 300 thousandths. Is he correct? Use a place value chart to explain your answer.

5. Alaska has a land area of about 1,700,000 square kilometers. Florida has a land area $\frac{1}{10}$ the size of Alaska. What is the land area of Florida? Explain how you found your answer.

Reason abstractly using place value understanding to relate adjacent base ten units from millions to thousandths.

1. Write the following in exponential form.

   a. $10 \times 10 \times 10 = \underline{\quad 10^3 \quad}$

   10 is a factor 3 times, so the exponent is 3. I can read this as, "ten to the third power."

   b. $1,000 \times 10 = \underline{\quad 10^4 \quad}$

   $1,000 = 10 \times 10 \times 10$, so this expression uses 10 as a factor 4 times. The exponent is 4.

   c. $100,000 = \underline{\quad 10^5 \quad}$

   d. $100 = \underline{\quad 10^2 \quad}$

   I recognize a pattern. 100 has 2 zeros. Therefore, the exponent is 2. One hundred equals 10 to the 2nd power.

2. Write the following in standard form.

   a. $6 \times 10^3 = \underline{\quad 6,000 \quad}$

   10^3 is equal to 1,000. 6 times 1 thousand is 6 thousand.

   b. $60.43 \times 10^4 = \underline{604,300}$

   The exponent 4 tells me how many places each digit will shift to the left.

   c. $643 \div 10^3 = \underline{\quad 0.643 \quad}$

   d. $6.4 \div 10^2 = \underline{\quad 0.064 \quad}$

   The exponent 2 tells me how many places each digit will shift to the right.

3. Complete the patterns.

   a. $0.06 \qquad 0.6 \qquad \underline{\quad 6 \quad} \qquad 60 \qquad \underline{\quad 600 \quad} \qquad \underline{\quad 6,000 \quad}$

   6 tenths is larger than 6 hundredths. Each number in the pattern is 10 times larger than the previous number.

   b. $\underline{92,100} \qquad 9,210 \qquad \underline{\quad 921 \quad} \qquad 92.1 \qquad 9.21 \qquad \underline{\quad 0.921 \quad}$

   The numbers are getting smaller in this pattern.

   The digits have each shifted 1 place to the right. The pattern in this sequence is "divide by 10^1."

EUREKA MATH

Lesson 3:   Use exponents to name place value units, and explain patterns in the placement of the decimal point.

© 2018 Great Minds®. eureka-math.org

13

Name _Adrienne landry_   Date _____

1. Write the following in exponential form (e.g., 100 = 10²).

    a. 1000 = 10^8

    b. 10 × 10 = 10^2

    c. 100,000 = 10^5

    d. 100 × 10 = 10^3

    e. 1,000,000 = 10^6

    f. 10,000 × 10 = 10^5

2. Write the following in standard form (e.g., 4 × 10² = 400).

    a. 4 × 10³ = 4,000

    b. 64 × 10⁴ = 640000

    c. 5,300 ÷ 10² = 5,300   10,000

    d. 5,300,000 ÷ 10³ = _____

    e. 6.072 × 10³ = 6072

    f. 60.72 × 10⁴ = 607,200
       10,000

    g. 948 ÷ 10³ = _____

    h. 9.4 ÷ 10² = _____

3. Complete the patterns.

    a. 0.02   0.2   _____   20   _____   _____

    b. 3,400,000   34,000   _____   3.4   _____

    c. _____   8,570   _____   85.7   8.57   _____

    d. 444   4440   44,400   _____   _____   _____

    e. _____   9.5   950   95,000   _____   _____

Lesson 3:   Use exponents to name place value units, and explain patterns in the placement of the decimal point.

15

© 2018 Great Minds®. eureka-math.org

4. After a lesson on exponents, Tia went home and said to her mom, "I learned that 10^4 is the same as 40,000." She has made a mistake in her thinking. Use words, numbers, or a place value chart to help Tia correct her mistake.

5. Solve $247 \div 10^2$ and 247×10^2.

   a. What is different about the two answers? Use words, numbers, or pictures to explain how the digits shift.

   b. Based on the answers from the pair of expressions above, solve $247 \div 10^3$ and 247×10^3.

Lesson 3: Use exponents to name place value units, and explain patterns in the placement of the decimal point.

© 2018 Great Minds®. eureka-math.org

1. Convert and write an equation with an exponent.

    In the first 2 problems, I am converting a *larger* unit to a *smaller* unit. Therefore, I need to multiply to find the equivalent length.

    > 1 meter is equal to 100 centimeters.

    a. 4 meters to centimeters   __4__ m = __400__ cm

    $$4 \times 10^2 = 400$$

    > 1 meter is equal to 1,000 millimeters.

    b. 2.8 meters to millimeters   __2.8__ m = __2,800__ mm

    $$2.8 \times 10^3 = 2,800$$

2. Convert using an equation with an exponent.

    In these 2 problems, I am converting a *smaller* unit to a *larger* unit. Therefore, I need to divide to find the equivalent length.

    > There are 100 centimeters in 1 meter.

    a. 87 centimeters to meters   __87__ cm = __0.87__ m

    $$87 \div 10^2 = 0.87$$

    > There are 1,000 millimeters in 1 meter.

    b. 9 millimeters to meters   __9__ mm = __0.009__ m

    $$9 \div 10^3 = 0.009$$

3. The height of a cellphone is 13 cm. Express this measurement in meters. Explain your thinking. Include an equation with an exponent in your explanation.

    $$13 \text{ cm} = 0.13 \text{ m}$$

    > In order to rename smaller units as larger units, I'll need to divide.

    Since 1 meter is equal to 100 centimeters, I divided the number of centimeters by 100.

    $$13 \div 10^2 = 0.13$$

    > I need to include an equation with an exponent, so I'll express 100 as 10^2.

$33 \div 6 = 5$ and there are three left over

OOOOO OOOOO OOOOO
OOOOO OOOOO OOO

She will
have 5 groups
of six

She will
have three
stutents in the
remaining group

Mr. D

Name ___Adrienne_____   Date _____

1. Convert and write an equation with an exponent. Use your meter strip when it helps you.

    a.   2 meters to centimeters   2m = 200 cm $2 \times 10^2 = 200$

    b.   108 centimeters to meters   108 cm = _____ m   _____

    c.   2.49 meters to centimeters   _____ m = _____ cm   _____

    d.   50 centimeters to meters   _____ cm = _____ m   _____

    e.   6.3 meters to centimeters   _____ m = _____ cm   _____

    f.   7 centimeters to meters   _____ cm = _____ m   _____

    g.   In the space below, list the letters of the problems where smaller units are converted to larger units.

2. Convert using an equation with an exponent. Use your meter strip when it helps you.

    a.   4 meters to millimeters   _____ m = _____ mm   _____

    b.   1.7 meters to millimeters   _____ m = _____ mm   _____

    c.   1,050 millimeters to meters _____ mm = _____ m   _____

    d.   65 millimeters to meters   _____ mm = _____ m   _____

    e.   4.92 meters to millimeters   _____ m = _____ mm   _____

    f.   3 millimeters to meters   _____ mm = _____ m   _____

    g.   In the space below, list the letters of the problems where larger units are converted to smaller units.

Lesson 4:   Use exponents to denote powers of 10 with application to metric
             conversions.

© 2018 Great Minds®. eureka-math.org

19

3. Read each aloud as you write the equivalent measures. Write an equation with an exponent you might use to convert.

a. 2.638 m = _____ mm $\underline{\quad 2.638 \times 10^3 = 2,638 \quad}$

b. 7 cm = _____ m   _____

c. 39 mm = _____ m   _____

d. 0.08 m = _____ mm   _____

e. 0.005 m = _____ cm   _____

4. Yi Ting's height is 1.49 m. Express this measurement in millimeters. Explain your thinking. Include an equation with an exponent in your explanation.

5. A ladybug's length measures 2 cm. Express this measurement in meters. Explain your thinking. Include an equation with an exponent in your explanation.

6. The length of a sticky note measures 77 millimeters. Express this length in meters. Explain your thinking. Include an equation with an exponent in your explanation.

Lesson 4:   Use exponents to denote powers of 10 with application to metric conversions.

© 2018 Great Minds®. eureka-math.org

EUREKA MATH®

1. Express as decimal numerals.

   a. Eight and three hundred fifty-two thousandths

      8.352

   b. $\dfrac{6}{100}$

      0.06

   c. $5\dfrac{132}{1000}$

      5.132

   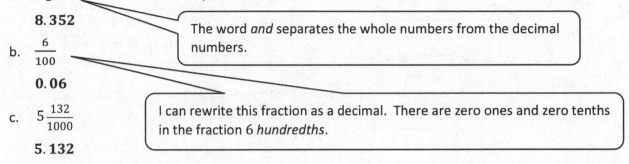

   > The word *and* separates the whole numbers from the decimal numbers.

   > I can rewrite this fraction as a decimal. There are zero ones and zero tenths in the fraction 6 *hundredths*.

2. Express in words.

   a. 0.034

      Thirty-four thousandths

   b. 73.29

      Seventy-three and twenty-nine hundredths

   > The word *and* separates the whole numbers from the decimal numbers.

3. Write the number in expanded form using decimals and fractions.

   303.084

   $$3 \times 100 + 3 \times 1 + 8 \times 0.01 + 4 \times 0.001$$

   $$3 \times 100 + 3 \times 1 + 8 \times \dfrac{1}{100} + 4 \times \dfrac{1}{1000}$$

   > This expanded form uses decimals. 8 hundredths is the same as 8 units of 1 hundredth or (8×0.01).

   > This expanded form uses fractions.
   > $\dfrac{1}{1000} = 0.001$
   > Both are read as one thousandths.

4. Write a decimal for each of the following.

a. $4 \times 100 + 5 \times 1 + 2 \times \frac{1}{10} + 8 \times \frac{1}{1000}$

   405.208

   There are 0 tens and 0 hundredths in expanded form, so I wrote 0 tens and 0 hundredths in standard form too.

b. $9 \times 1 + 9 \times 0.1 + 3 \times 0.01 + 6 \times 0.001$

   9.936

   3×0.01 is 3 units of 1 hundredth, which I can write as a 3 in the hundredths place.

Lesson 5:   Name decimal fractions in expanded, unit, and word forms by applying place value reasoning.

Name _Adrienne_____   Date _____

1. Express as decimal numerals. The first one is done for you.

standard form

a.	Five thousandths	0.005
b.	Thirty-five thousandths	0.035
c.	Nine and two hundred thirty-five thousandths	9.255
d.	Eight hundred and five thousandths	800.005
e.	$\frac{8}{1000}$	8.000
f.	$\frac{28}{1000}$	28.00
g.	$7\frac{528}{1000}$	
h.	$300\frac{502}{1000}$	

2. Express each of the following values in words.

   a. 0.008 ___eight thousanths_____

   b. 15.062 ___fifteen and sixty two Thousanths_____

   c. 607.409 ___Sixhundred seven and fourhundred nine thousants

3. Write the number on a place value chart. Then, write it in expanded form using fractions or decimals to express the decimal place value units. The first one is done for you.

   a. 27.346

Tens	Ones	●	Tenths	Hundredths	Thousandths
2	7		3	4	6

   $27.346 = 2 \times 10 + 7 \times 1 + 3 \times \left(\frac{1}{10}\right) + 4 \times \left(\frac{1}{100}\right) + 6 \times \left(\frac{1}{1000}\right)$ or

   $27.346 = 2 \times 10 + 7 \times 1 + 3 \times 0.1 + 4 \times 0.01 + 6 \times 0.001$

Lesson 5:   Name decimal fractions in expanded, unit, and word forms by
applying place value reasoning.

© 2018 Great Minds®. eureka-math.org

23

b. 0.362

c. 49.564

4. Write a decimal for each of the following. Use a place value chart to help, if necessary.

a. $3 \times 10 + 5 \times 1 + 2 \times \left(\frac{1}{10}\right) + 7 \times \left(\frac{1}{100}\right) + 6 \times \left(\frac{1}{1000}\right)$

b. $9 \times 100 + 2 \times 10 + 3 \times 0.1 + 7 \times 0.001$

c. $5 \times 1000 + 4 \times 100 + 8 \times 1 + 6 \times \left(\frac{1}{100}\right) + 5 \times \left(\frac{1}{1000}\right)$

5. At the beginning of a lesson, a piece of chalk is 4.875 inches long. At the end of the lesson, it is 3.125 inches long. Write the two amounts in expanded form using fractions.

a. At the beginning of the lesson:

b. At the end of the lesson:

6. Mrs. Herman asked the class to write an expanded form for 412.638. Nancy wrote the expanded form using fractions, and Charles wrote the expanded form using decimals. Write their responses.

Lesson 5:   Name decimal fractions in expanded, unit, and word forms by
applying place value reasoning.

© 2018 Great Minds®. eureka-math.org

Thousandths	
Hundredths	
Tenths	
● Ones	
Tens	
Hundreds	
Thousands	

thousands through thousandths place value chart

Lesson 5:   Name decimal fractions in expanded, unit, and word forms by
applying place value reasoning.

25

© 2018 Great Minds®. eureka-math.org

1. Show the numbers on the place value chart using digits. Use >, <, or = to compare.

   43.554 __>__ 43.545

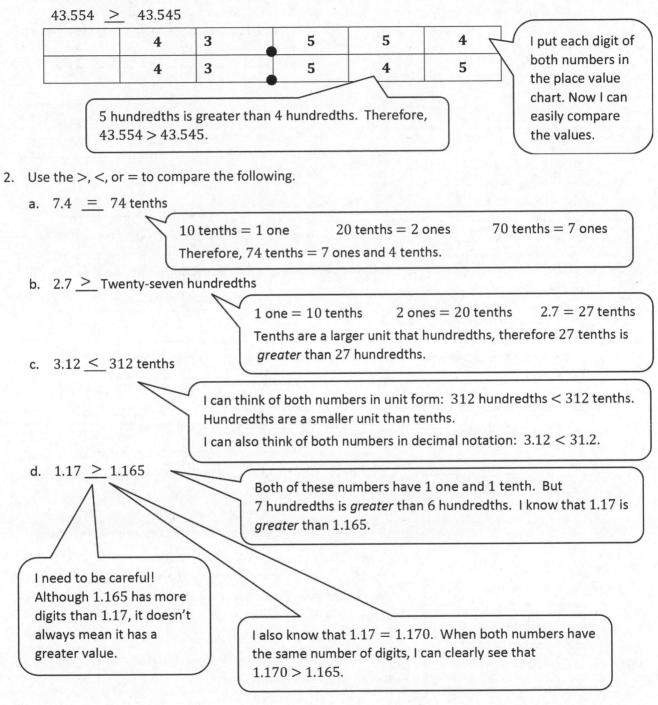

	4	3	•	5	5	4
	4	3	•	5	4	5

   5 hundredths is greater than 4 hundredths. Therefore, 43.554 > 43.545.

   I put each digit of both numbers in the place value chart. Now I can easily compare the values.

2. Use the >, <, or = to compare the following.

   a. 7.4 __=__ 74 tenths

   10 tenths = 1 one   20 tenths = 2 ones   70 tenths = 7 ones
   Therefore, 74 tenths = 7 ones and 4 tenths.

   b. 2.7 __>__ Twenty-seven hundredths

   1 one = 10 tenths   2 ones = 20 tenths   2.7 = 27 tenths
   Tenths are a larger unit that hundredths, therefore 27 tenths is *greater* than 27 hundredths.

   c. 3.12 __<__ 312 tenths

   I can think of both numbers in unit form: 312 hundredths < 312 tenths. Hundredths are a smaller unit than tenths.
   I can also think of both numbers in decimal notation: 3.12 < 31.2.

   d. 1.17 __>__ 1.165

   Both of these numbers have 1 one and 1 tenth. But 7 hundredths is *greater* than 6 hundredths. I know that 1.17 is *greater* than 1.165.

   I need to be careful! Although 1.165 has more digits than 1.17, it doesn't always mean it has a greater value.

   I also know that 1.17 = 1.170. When both numbers have the same number of digits, I can clearly see that 1.170 > 1.165.

Lesson 6:   Compare decimal fractions to the thousandths using like units, and express comparisons with >, <, =.

27

© 2018 Great Minds®. eureka-math.org

3. Arrange the numbers in *increasing* order.

8.719   8.79   8.7   8.179

8.179,   8.7,   8.719,   8.79

Increasing order means I need to list the numbers from *least* to *greatest*.

To make comparing easier, I'm going to use a place value chart.

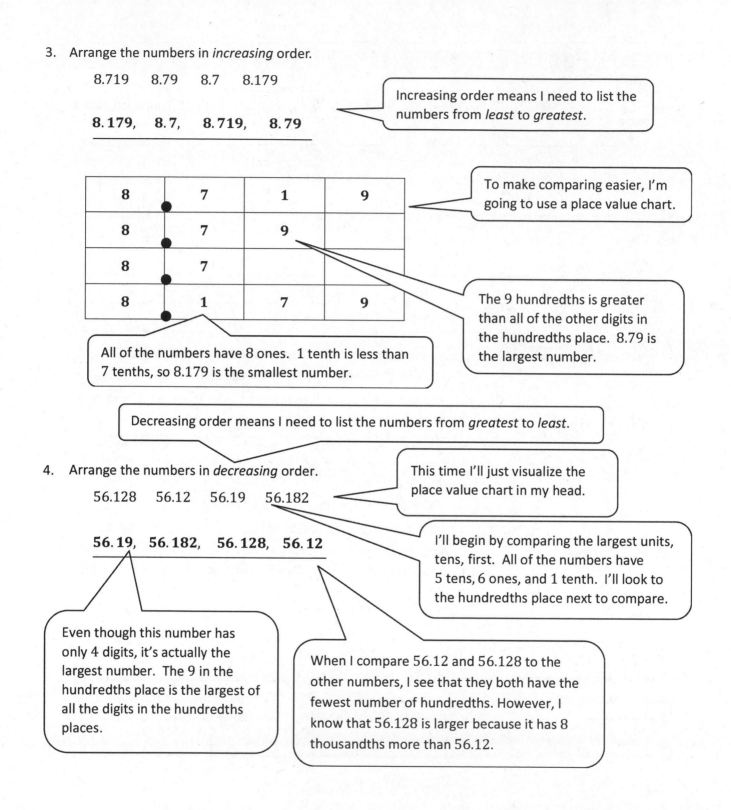

The 9 hundredths is greater than all of the other digits in the hundredths place. 8.79 is the largest number.

All of the numbers have 8 ones. 1 tenth is less than 7 tenths, so 8.179 is the smallest number.

Decreasing order means I need to list the numbers from *greatest* to *least*.

4. Arrange the numbers in *decreasing* order.

56.128   56.12   56.19   56.182

56.19,   56.182,   56.128,   56.12

This time I'll just visualize the place value chart in my head.

I'll begin by comparing the largest units, tens, first. All of the numbers have 5 tens, 6 ones, and 1 tenth. I'll look to the hundredths place next to compare.

Even though this number has only 4 digits, it's actually the largest number. The 9 in the hundredths place is the largest of all the digits in the hundredths places.

When I compare 56.12 and 56.128 to the other numbers, I see that they both have the fewest number of hundredths. However, I know that 56.128 is larger because it has 8 thousandths more than 56.12.

Lesson 6:   Compare decimal fractions to the thousandths using like units, and express comparisons with >, <, =.

© 2018 Great Minds®. eureka-math.org

Name _____   Date _____

1. Use >, <, or = to compare the following.

a. 16.45	$\bigcirc$	16.454
b. 0.83	$\bigcirc$	$\dfrac{83}{100}$
c. $\dfrac{205}{1000}$	$\bigcirc$	0.205
d. 95.045	$\bigcirc$	95.545
e. 419.10	$\bigcirc$	419.099
f. Five ones and eight tenths	$\bigcirc$	Fifty-eight tenths
g. Thirty-six and nine thousandths	$\bigcirc$	Four tens
h. One hundred four and twelve hundredths	$\bigcirc$	One hundred four and two thousandths
i. One hundred fifty-eight thousandths	$\bigcirc$	0.58
j. 703.005	$\bigcirc$	Seven hundred three and five hundredths

2. Arrange the numbers in increasing order.

a. 8.08   8.081   8.09   8.008

_____

b. 14.204   14.200   14.240   14.210

_____

Lesson 6:   Compare decimal fractions to the thousandths using like units, and
express comparisons with >, <, =.

© 2018 Great Minds®. eureka-math.org

29

3. Arrange the numbers in decreasing order.

   a. 8.508   8.58   7.5   7.058

   _____

   b. 439.216   439.126   439.612   439.261

   _____

4. James measured his hand. It was 0.17 meter. Jennifer measured her hand. It was 0.165 meter. Whose hand is bigger? How do you know?

5. In a paper airplane contest, Marcel's plane travels 3.345 meters. Salvador's plane travels 3.35 meters. Jennifer's plane travels 3.3 meters. Based on the measurements, whose plane traveled the farthest distance? Whose plane traveled the shortest distance? Explain your reasoning using a place value chart.

Lesson 6:   Compare decimal fractions to the thousandths using like units, and express comparisons with >, <, =.

Round to the given place value. Label the number lines to show your work. Circle the rounded number. Use a place value chart to show your decompositions for each.

1. 3.27

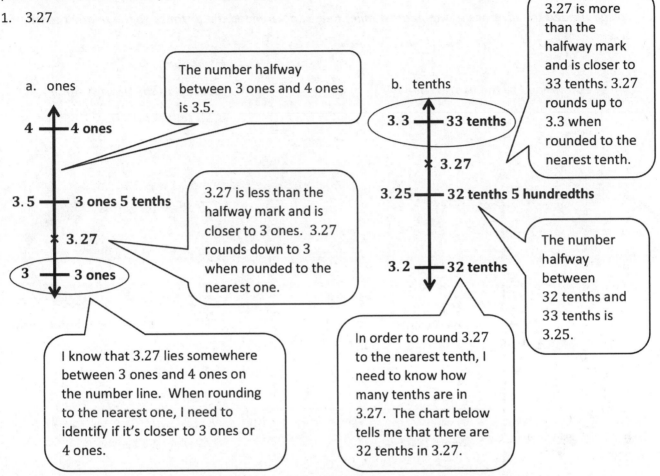

a. ones

4 — 4 ones

The number halfway between 3 ones and 4 ones is 3.5.

3.5 — 3 ones 5 tenths

✗ 3.27

3 — 3 ones

3.27 is less than the halfway mark and is closer to 3 ones. 3.27 rounds down to 3 when rounded to the nearest one.

I know that 3.27 lies somewhere between 3 ones and 4 ones on the number line. When rounding to the nearest one, I need to identify if it's closer to 3 ones or 4 ones.

b. tenths

3.3 — 33 tenths

✗ 3.27

3.25 — 32 tenths 5 hundredths

3.2 — 32 tenths

3.27 is more than the halfway mark and is closer to 33 tenths. 3.27 rounds up to 3.3 when rounded to the nearest tenth.

The number halfway between 32 tenths and 33 tenths is 3.25.

In order to round 3.27 to the nearest tenth, I need to know how many tenths are in 3.27. The chart below tells me that there are 32 tenths in 3.27.

ones	tenths	hundredths
3	2	7
	32	7
		327

I can think of 3.27 in several ways. I can say it is 3 ones + 2 tenths + 7 hundredths. I can also think of it as 32 tenths + 7 hundredths or 327 hundredths.

Lesson 7: Round a given decimal to any place using place value understanding and the vertical number line.

31

2. Rosie's pedometer said she walked 1.46 miles. She rounded her distance to 1 mile, and her brother, Isaac, rounded her distance to 1.5 miles. They are both right. Why?

Rosie rounded the distance to the nearest mile, and Isaac rounded the distance to the nearest tenth of a mile.

1.46 rounded to the nearest one is 1.   *1.46 rounded to the nearest tenth is 15 tenths or 1.5.*

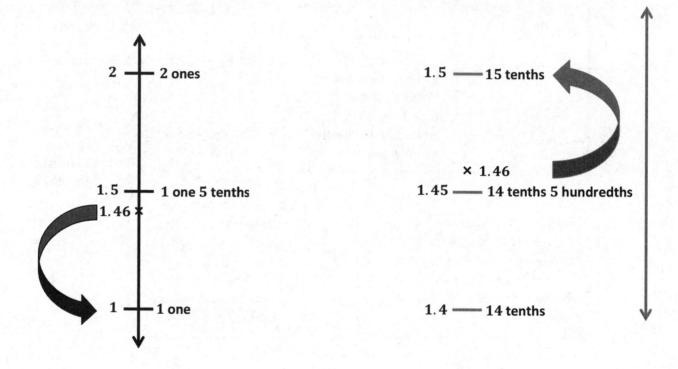

Lesson 7:   Round a given decimal to any place using place value understanding and the vertical number line.

© 2018 Great Minds®. eureka-math.org

Name _____ Date _____

Fill in the table, and then round to the given place. Label the number lines to show your work. Circle the rounded number.

1. 4.3

   a. Hundredths   b. Tenths   c. Ones

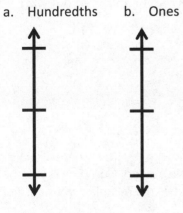

Tens	Ones	Tenths	Hundredths	Thousandths

2. 225.286

   a. Hundredths   b. Ones   c. Tens

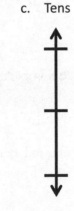

Tens	Ones	Tenths	Hundredths	Thousandths

Lesson 7:   Round a given decimal to any place using place value understanding
             and the vertical number line.

© 2018 Great Minds®. eureka-math.org

33

3. 8.984

Tens	Ones	Tenths	Hundredths	Thousandths

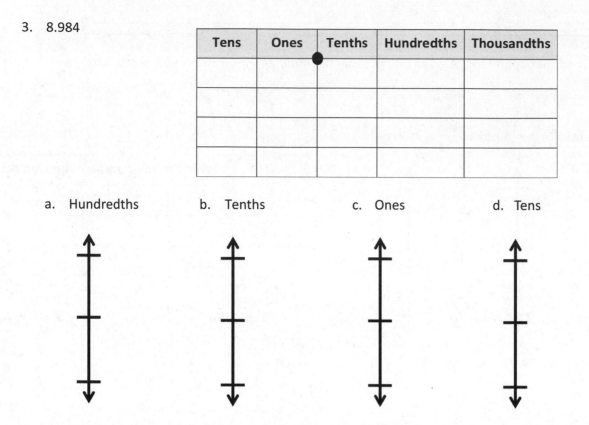

a. Hundredths   b. Tenths   c. Ones   d. Tens

4. On a Major League Baseball diamond, the distance from the pitcher's mound to home plate is 18.386 meters.

    a. Round this number to the nearest hundredth of a meter. Use a number line to show your work.

    b. How many centimeters is it from the pitcher's mound to home plate?

5. Jules reads that 1 pint is equivalent to 0.473 liters. He asks his teacher how many liters there are in a pint. His teacher responds that there are about 0.47 liters in a pint. He asks his parents, and they say there are about 0.5 liters in a pint. Jules says they are both correct. How can that be true? Explain your answer.

Lesson 7:   Round a given decimal to any place using place value understanding and the vertical number line.

© 2018 Great Minds®. eureka-math.org

1. Round the quantity to the given place value. Draw number lines to explain your thinking. Circle the rounded value on the number line.

Round 23.245 to the nearest tenth and hundredth.

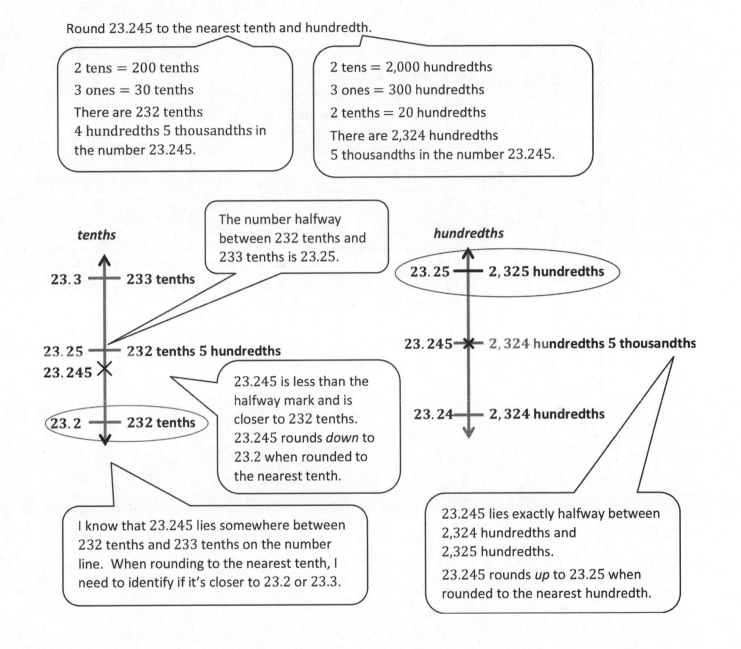

2 tens = 200 tenths

3 ones = 30 tenths

There are 232 tenths 4 hundredths 5 thousandths in the number 23.245.

2 tens = 2,000 hundredths

3 ones = 300 hundredths

2 tenths = 20 hundredths

There are 2,324 hundredths 5 thousandths in the number 23.245.

tenths

23.3 ─┼─ 233 tenths

The number halfway between 232 tenths and 233 tenths is 23.25.

23.25 ─┼─ 232 tenths 5 hundredths

23.245 ─✕─

23.245 is less than the halfway mark and is closer to 232 tenths. 23.245 rounds *down* to 23.2 when rounded to the nearest tenth.

(23.2) ─┼─ 232 tenths

I know that 23.245 lies somewhere between 232 tenths and 233 tenths on the number line. When rounding to the nearest tenth, I need to identify if it's closer to 23.2 or 23.3.

hundredths

23.25 ─┼─ 2,325 hundredths

23.245 ─✕─ 2,324 hundredths 5 thousandths

23.24 ─┼─ 2,324 hundredths

23.245 lies exactly halfway between 2,324 hundredths and 2,325 hundredths.

23.245 rounds *up* to 23.25 when rounded to the nearest hundredth.

Lesson 8:   Round a given decimal to any place using place value understanding and the vertical number line.

© 2018 Great Minds®. eureka-math.org

35

2. A decimal number has two digits to the right of its decimal point. If we round it to the nearest tenth, the result is 28.7. What is the maximum possible value of this decimal? Use words and the number line to explain your reasoning.

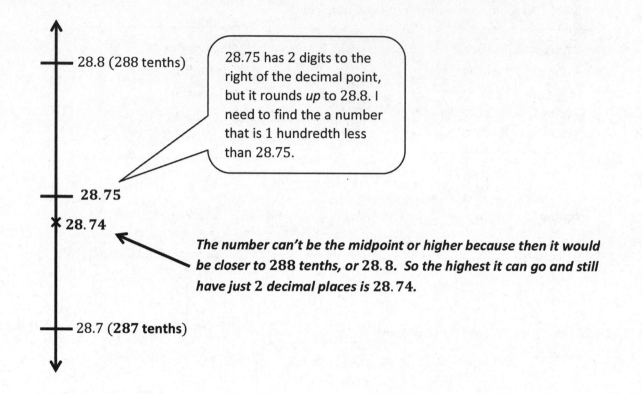

28.75 has 2 digits to the right of the decimal point, but it rounds *up* to 28.8. I need to find the a number that is 1 hundredth less than 28.75.

The number can't be the midpoint or higher because then it would be closer to 288 tenths, or 28.8. So the highest it can go and still have just 2 decimal places is 28.74.

    Lesson 8:   Round a given decimal to any place using place value understanding and the vertical number line.

© 2018 Great Minds®. eureka-math.org

Name _____ Date _____

1. Write the decomposition that helps you, and then round to the given place value. Draw number lines to explain your thinking. Circle the rounded value on each number line.

   a. 43.586 to the nearest tenth, hundredth, and one.

   b. 243.875 to nearest tenth, hundredth, ten, and hundred.

2. A trip from New York City to Seattle is 2,852.1 miles. A family wants to make the drive in 10 days, driving the same number of miles each day. About how many miles will they drive each day? Round your answer to the nearest tenth of a mile.

Lesson 8: Round a given decimal to any place using place value understanding and the vertical number line.

© 2018 Great Minds®. eureka-math.org

37

3. A decimal number has two digits to the right of its decimal point. If we round it to the nearest tenth, the result is 18.6.

   a. What is the maximum possible value of this number? Use words and the number line to explain your reasoning. Include the midpoint on your number line.

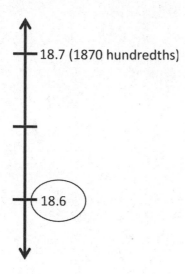

   b. What is the minimum possible value of this decimal? Use words, pictures, or numbers to explain your reasoning.

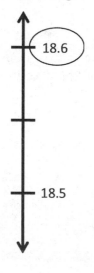

Lesson 8:   Round a given decimal to any place using place value understanding and the vertical number line.

Note: Adding decimals is just like adding whole numbers—combine like units. Study the examples below:

2 apples + 3 apples = 5 apples

2 ones + 3 ones = 5 ones

2 tens + 3 tens = 5 tens = 50

2 hundredths + 3 hundredths = 5 hundredths = 0.05

1. Solve.

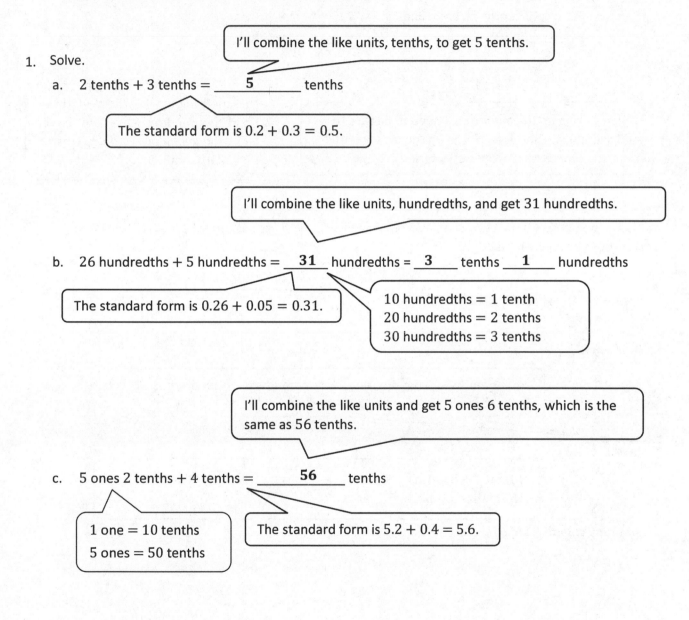

I'll combine the like units, tenths, to get 5 tenths.

a. 2 tenths + 3 tenths = _____**5**_____ tenths

The standard form is 0.2 + 0.3 = 0.5.

I'll combine the like units, hundredths, and get 31 hundredths.

b. 26 hundredths + 5 hundredths = __**31**__ hundredths = __**3**__ tenths __**1**__ hundredths

The standard form is 0.26 + 0.05 = 0.31.

10 hundredths = 1 tenth
20 hundredths = 2 tenths
30 hundredths = 3 tenths

I'll combine the like units and get 5 ones 6 tenths, which is the same as 56 tenths.

c. 5 ones 2 tenths + 4 tenths = _____**56**_____ tenths

1 one = 10 tenths
5 ones = 50 tenths

The standard form is 5.2 + 0.4 = 5.6.

2. Solve using the standard algorithm.

a. $0.3 + 0.91 =$ __**1.21**__

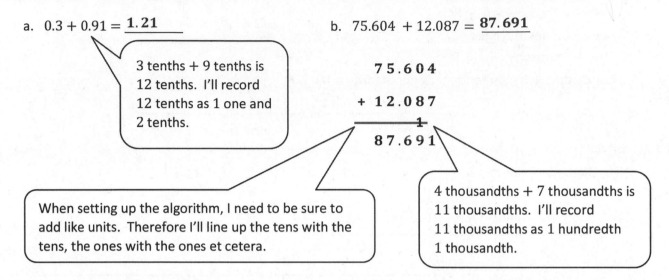

3 tenths + 9 tenths is
12 tenths. I'll record
12 tenths as 1 one and
2 tenths.

b. $75.604 + 12.087 =$ **87.691**

$$
\begin{array}{r}
7\,5.6\,0\,4 \\
+\ 1\,2.0\,8\,7 \\
\hline
1 \\
8\,7.6\,9\,1
\end{array}
$$

When setting up the algorithm, I need to be sure to add like units. Therefore I'll line up the tens with the tens, the ones with the ones et cetera.

4 thousandths + 7 thousandths is 11 thousandths. I'll record 11 thousandths as 1 hundredth 1 thousandth.

3. Anthony spends $6.49 on a book. He also buys a pencil for $2.87 and an eraser for $1.15. How much money does he spend altogether?

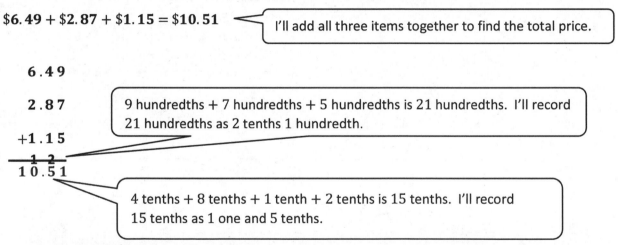

$\$6.49 + \$2.87 + \$1.15 = \$10.51$

I'll add all three items together to find the total price.

$$
\begin{array}{r}
6.4\,9 \\
2.8\,7 \\
+\,1.1\,5 \\
\hline
1\ 2 \\
1\,0.5\,1
\end{array}
$$

9 hundredths + 7 hundredths + 5 hundredths is 21 hundredths. I'll record 21 hundredths as 2 tenths 1 hundredth.

4 tenths + 8 tenths + 1 tenth + 2 tenths is 15 tenths. I'll record 15 tenths as 1 one and 5 tenths.

Anthony spends $10.51.

Lesson 9:   Add decimals using place value strategies, and relate those strategies to a written method.

EUREKA
MATH®

Note: Subtracting decimals is just like subtracting whole numbers—subtract like units. Study the examples below.

5 apples − 1 apple = 4 apples
5 ones − 1 one = 4 ones
5 tens − 1 ten = 4 tens
5 hundredths − 1 hundredth = 4 hundredths

1. Subtract.

> I'll subtract the like units, tenths, to get 3 tenths.

   a. 7 tenths −4 tenths = ____3____ tenths

> The standard form is 0.7 − 0.4 = 0.3.

> I'll look at the units carefully.
> A *hundred* is different than a *hundredth*.

> I'll subtract 3 hundredths from 8 hundredths, and get 5 hundredths.

   b. 4 hundreds 8 hundredths −3 hundredths = ____4____ hundreds ____5____ hundredths

> The standard form is 400.08 − 0.03 = 400.05.

> 1.7 is the same as 1.70.

2. Solve 1.7 − 0.09 using the standard algorithm.

> When setting up the algorithm, I need to be sure to subtract like units. Therefore, I'll line up the ones with the ones, the tenths with the tenths, etc.

> There are 0 hundredths, so I can't subtract 9 hundredths. I'll rename 7 tenths as 6 tenths 10 hundredths.

> 10 hundredths minus 9 hundredths is equal to 1 hundredth.

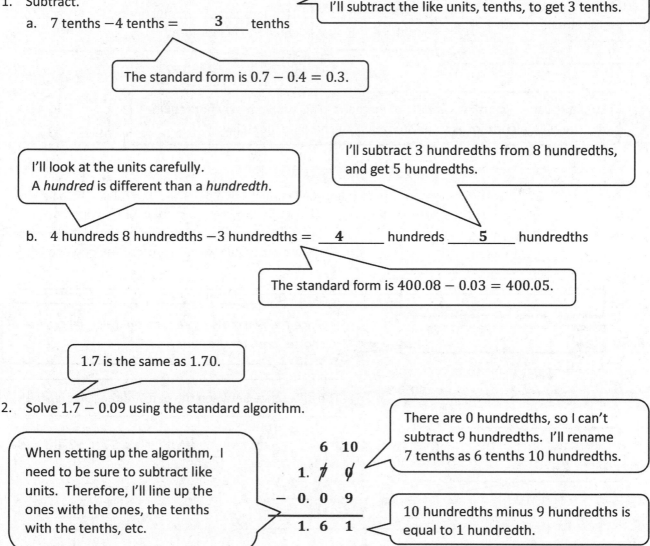

$$
\begin{array}{r}
\overset{6}{\cancel{1}}.\overset{10}{\cancel{7}}\ \cancel{0} \\
-\ 0.\ 0\ 9 \\
\hline
1.\ 6\ 1
\end{array}
$$

6 ones 3 tenths = 6.3 = 6.30
58 hundredths = 0.58

3. Solve 6 ones 3 tenths −58 hundredths.

There are 0 hundredths, so I can't subtract
8 hundredths. I'll rename 3 tenths as 2
tenths 10 hundredths.

I'll rename 6 ones as 5 ones
10 tenths. 10 tenths, plus the
2 tenths already there, makes
12 tenths.

$$
\begin{array}{r}
5\ \ \ 12\ \ 10 \\
\cancel{6}.\ \cancel{3}\ \ \cancel{0} \\
-\ \ 0.\ 5\ \ 8 \\
\hline
5.\ 7\ \ 2
\end{array}
$$

10 hundredths minus 8 hundredths is
equal to 2 hundredths.

Students can solve using a variety of methods. This problem may not require
the standard algorithm as some students can compute mentally.

4. A pen costs $2.57. It costs $0.49 more than a ruler. Kayla bought two pens and one ruler. She paid with
a ten-dollar bill. How much change does Kayla get? Use a tape diagram to show your thinking.

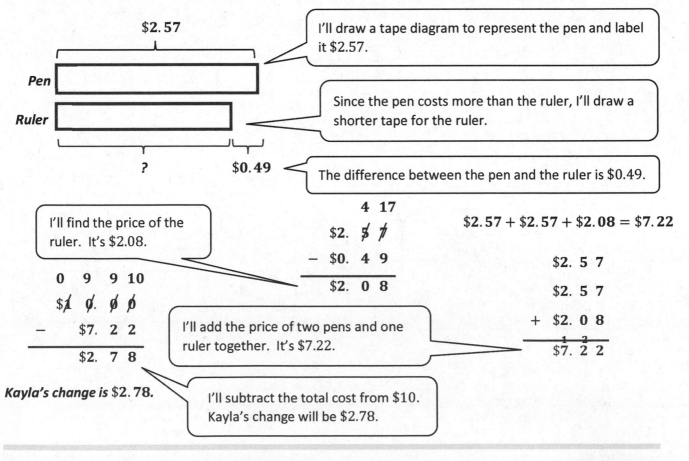

$2.57

I'll draw a tape diagram to represent the pen and label
it $2.57.

Pen

Ruler

Since the pen costs more than the ruler, I'll draw a
shorter tape for the ruler.

? $0.49

The difference between the pen and the ruler is $0.49.

I'll find the price of the
ruler. It's $2.08.

$$
\begin{array}{r}
4\ \ 17 \\
\$2.\ \cancel{5}\ \cancel{7} \\
-\ \$0.\ 4\ 9 \\
\hline
\$2.\ 0\ 8
\end{array}
$$

$2.57 + $2.57 + $2.08 = $7.22

$$
\begin{array}{r}
0\ \ \ 9\ \ 9\ \ 10 \\
\$\cancel{1}\ \cancel{0}.\ \cancel{0}\ \cancel{0} \\
-\ \ \ \ \$7.\ 2\ 2 \\
\hline
\$2.\ 7\ 8
\end{array}
$$

I'll add the price of two pens and one
ruler together. It's $7.22.

$$
\begin{array}{r}
\$2.\ 5\ 7 \\
\$2.\ 5\ 7 \\
+\ \$2.\ 0\ 8 \\
\hline
\$7.\ 2\ 2
\end{array}
$$

Kayla's change is $2.78.

I'll subtract the total cost from $10.
Kayla's change will be $2.78.

Lesson 10:   Subtract decimals using place value strategies, and relate those
strategies to a written method.

© 2018 Great Minds®. eureka-math.org

EUREKA
MATH®

Note: Encourage your child to use a variety of strategies when solving. The standard algorithm may not always be necessary for some students. Ask them about different ways to solve the problem. Below you'll find some alternate solution strategies that could be applied.

$$\$2.57 + \$2.57 + \$2.08 = \$7.22$$

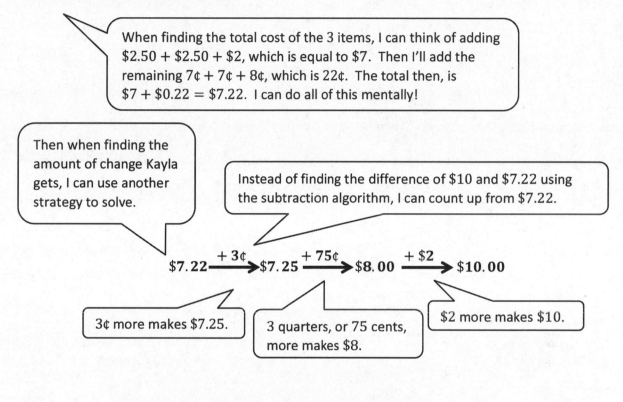

When finding the total cost of the 3 items, I can think of adding $2.50 + $2.50 + $2, which is equal to $7. Then I'll add the remaining 7¢ + 7¢ + 8¢, which is 22¢. The total then, is $7 + $0.22 = $7.22. I can do all of this mentally!

Then when finding the amount of change Kayla gets, I can use another strategy to solve.

Instead of finding the difference of $10 and $7.22 using the subtraction algorithm, I can count up from $7.22.

$$\$7.22 \xrightarrow{+ 3¢} \$7.25 \xrightarrow{+ 75¢} \$8.00 \xrightarrow{+ \$2} \$10.00$$

3¢ more makes $7.25.

3 quarters, or 75 cents, more makes $8.

$2 more makes $10.

2 dollars, 3 quarters, and 3 pennies is $2.78. That's what Kayla gets back.

Kayla gets $2.78 *back in change.*

Lesson 10: Subtract decimals using place value strategies, and relate those strategies to a written method.

45

© 2018 Great Minds®. eureka-math.org

Name _____   Date _____

1. Subtract. You may use a place value chart.

    a. 9 tenths – 3 tenths = _____ tenths

    b. 9 ones 2 thousandths – 3 ones = _____ ones _____ thousandths

    c. 4 hundreds 6 hundredths – 3 hundredths = _____ hundreds _____ hundredths

    d. 56 thousandths – 23 thousandths = _____ thousandths = _____ hundredths _____ thousandths

2. Solve using the standard algorithm.

a. 1.8 – 0.9 = _____	b. 41.84 – 0.9 = _____	c. 341.84 – 21.92 = _____
d. 5.182 – 0.09 = _____	e. 50.416 – 4.25 = _____	f. 741 – 3.91 = _____

Lesson 10:   Subtract decimals using place value strategies, and relate those
strategies to a written method.

© 2018 Great Minds®. eureka-math.org

47

3. Solve.

a. 30 tens – 3 tens 3 tenths	b. 5 – 16 tenths	c. 24 tenths – 1 one 3 tenths
d. 6 ones 7 hundredths – 2.3	e. 8.246 – 5 hundredths	f. 5 ones 3 tenths – 0.53

4. Mr. House wrote *8 tenths minus 5 hundredths* on the board. Maggie said the answer is 3 hundredths because 8 minus 5 is 3. Is she correct? Explain.

5. A clipboard costs $2.23. It costs $0.58 more than a notebook. Lisa bought two clipboards and one notebook. She paid with a ten-dollar bill. How much change does Lisa get? Use a tape diagram to show your thinking.

     Lesson 10:   Subtract decimals using place value strategies, and relate those strategies to a written method.

© 2018 Great Minds®. eureka-math.org

Hundreds	Tens	Ones	•	Tenths	Hundredths	Thousandths

hundreds to thousandths place value chart (from Lesson 7)

Lesson 10: Subtract decimals using place value strategies, and relate those strategies to a written method.

© 2018 Great Minds®. eureka-math.org

49

1. Solve by drawing disks on a place value chart. Write an equation, and express the product in standard form.

a. 2 copies of 4 tenths

$= 2 \times 0.4$
$= 0.8$

2 copies means 2 groups. So, I'll multiply 2 times 4 tenths. The answer is 8 tenths, or 0.8.

I'll draw a place value chart to help me solve, and this dot is the decimal point.

Each dot represents 1 tenth, so I'll draw 2 groups of 4 tenths.

b. 3 times as much as 6 hundredths

$= 3 \times 0.06$
$= 0.18$

I'll multiply 3 times 6 hundredths. The answer is 18 hundredths, or 0.18.

I'll draw 3 groups of 6 hundredths.

I'll bundle 10 hundredths and exchange them for 1 tenth.

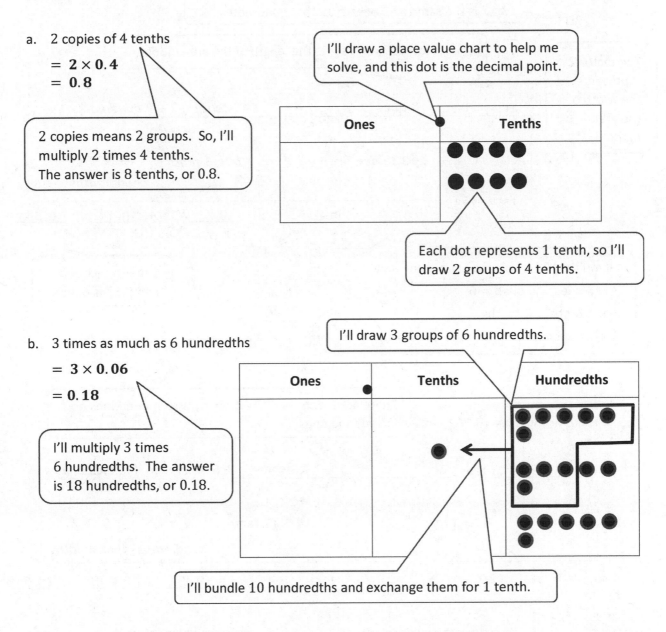

Lesson 11: Multiply a decimal fraction by single-digit whole numbers, relate to a written method through application of the area model and place value understanding, and explain the reasoning used.

© 2018 Great Minds®. eureka-math.org

51

2. Draw an area model, and find the sum of the partial products to evaluate each expression.

a. 2×3.17

3.17 is the same as 3 ones 1 tenth 7 hundredths.

The factor 3.17 represents the length of the area model.

The factor 2 represents the width of the area model.

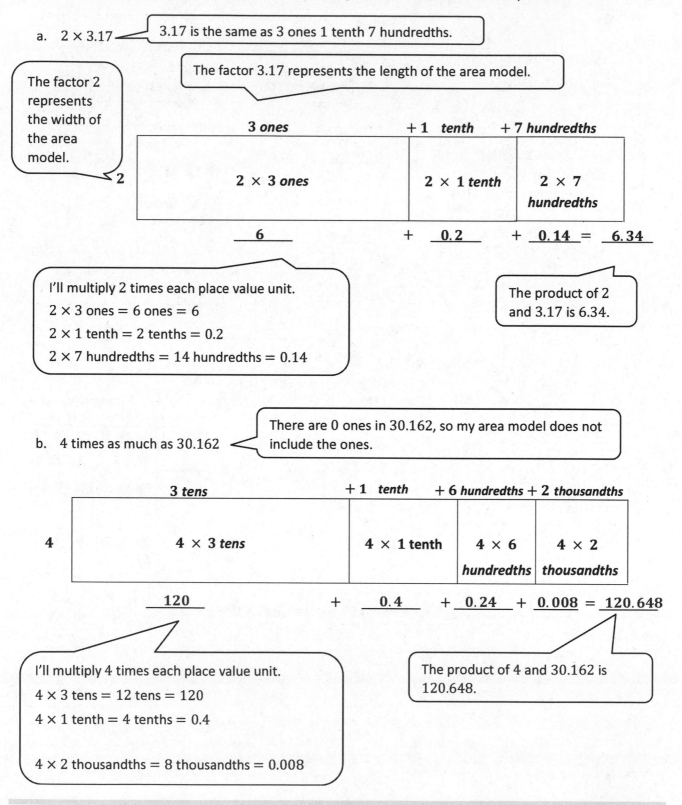

	3 ones	+ 1 tenth	+ 7 hundredths
2	2 × 3 ones	2 × 1 tenth	2 × 7 hundredths
	6	+ 0.2	+ 0.14 = 6.34

I'll multiply 2 times each place value unit.

2×3 ones = 6 ones = 6

2×1 tenth = 2 tenths = 0.2

2×7 hundredths = 14 hundredths = 0.14

The product of 2 and 3.17 is 6.34.

b. 4 times as much as 30.162

There are 0 ones in 30.162, so my area model does not include the ones.

	3 tens	+ 1 tenth	+ 6 hundredths	+ 2 thousandths
4	4 × 3 tens	4 × 1 tenth	4 × 6 hundredths	4 × 2 thousandths
	120	+ 0.4	+ 0.24	+ 0.008 = 120.648

I'll multiply 4 times each place value unit.

4×3 tens = 12 tens = 120

4×1 tenth = 4 tenths = 0.4

4×2 thousandths = 8 thousandths = 0.008

The product of 4 and 30.162 is 120.648.

Lesson 11: Multiply a decimal fraction by single-digit whole numbers, relate to a written method through application of the area model and place value understanding, and explain the reasoning used.

Name _____   Date _____

1. Solve by drawing disks on a place value chart. Write an equation, and express the product in standard form.

    a. 2 copies of 4 tenths

    b. 4 groups of 5 hundredths

    c. 4 times 7 tenths

    d. 3 times 5 hundredths

    e. 9 times as much as 7 tenths

    f. 6 thousandths times 8

2. Draw a model similar to the one pictured below. Find the sum of the partial products to evaluate each expression.

    a. 4 × 6.79

    6 ones   +   7 tenths   +   9 hundredths

    | 4 | 4 × 6 ones | 4 × 7 tenths | 4 × 9 hundredths |

    _____ + _____ + _____ = _____

Lesson 11:   Multiply a decimal fraction by single-digit whole numbers, relate to a written method through application of the area model and place value understanding, and explain the reasoning used.

© 2018 Great Minds®. eureka-math.org

53

b.   6 × 7.49

c.   9 copies of 3.65

d.   3 times 20.175

3.   Leanne multiplied 8 × 4.3 and got 32.24. Is Leanne correct? Use an area model to explain your answer.

4.   Anna buys groceries for her family. Hamburger meat is $3.38 per pound, sweet potatoes are $0.79 each, and hamburger rolls are $2.30 a bag. If Anna buys 3 pounds of meat, 5 sweet potatoes, and 1 bag of hamburger rolls, what will she pay in all for the groceries?

Lesson 11: Multiply a decimal fraction by single-digit whole numbers, relate to a written method through application of the area model and place value understanding, and explain the reasoning used.

1. Choose the reasonable product for each expression. Explain your thinking in the spaces below using words, pictures, or numbers.

a. 3.1×3 930   93   9.3   0.93

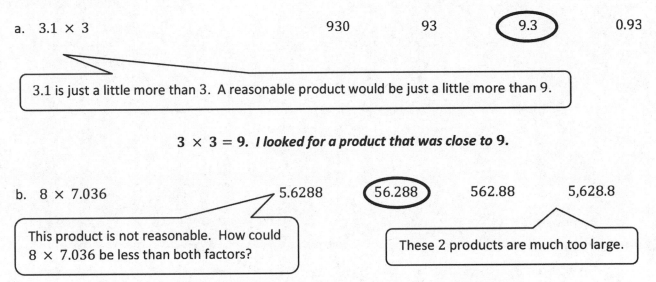

3.1 is just a little more than 3. A reasonable product would be just a little more than 9.

$3 \times 3 = 9$. *I looked for a product that was close to* **9.**

b. 8×7.036 5.6288   56.288   562.88   5,628.8

This product is not reasonable. How could 8×7.036 be less than both factors?

These 2 products are much too large.

$8 \times 7 = 56$. *I looked for a product that was close to* **56.**

2. Lenox weighs 9.2 kg. Her older brother is 3 times as heavy as Lenox. How much does her older brother weigh in kilograms?

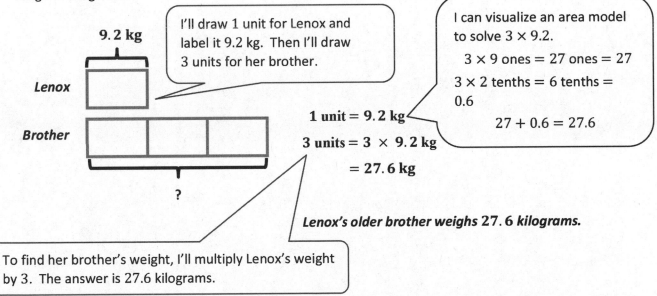

9.2 kg

I'll draw 1 unit for Lenox and label it 9.2 kg. Then I'll draw 3 units for her brother.

I can visualize an area model to solve 3×9.2.

3×9 ones $= 27$ ones $= 27$

3×2 tenths $= 6$ tenths $= 0.6$

$27 + 0.6 = 27.6$

Lenox

Brother

?

$1 \text{ unit} = 9.2 \text{ kg}$

$3 \text{ units} = 3 \times 9.2 \text{ kg}$

$= 27.6 \text{ kg}$

Lenox's older brother weighs **27.6** *kilograms.*

To find her brother's weight, I'll multiply Lenox's weight by 3. The answer is 27.6 kilograms.

Lesson 12:   Multiply a decimal fraction by single-digit whole numbers, including using estimation to confirm the placement of the decimal point.

55

© 2018 Great Minds®. eureka-math.org

Name _____ Date _____

1. Choose the reasonable product for each expression. Explain your thinking in the spaces below using words, pictures, or numbers.

   a.   2.1 × 3   0.63   6.3   63   630

   _____

   b.   4.27 × 6   2562   256.2   25.62   2.562

   _____

   c.   7 × 6.053   4237.1   423.71   42.371   4.2371

   _____

   d.   9 × 4.82   4.338   43.38   433.8   4338

Lesson 12:   Multiply a decimal fraction by single-digit whole numbers, including
               using estimation to confirm the placement of the decimal point.

57

© 2018 Great Minds®. eureka-math.org

2.   Yi Ting weighs 8.3 kg. Her older brother is 4 times as heavy as Yi Ting. How much does her older brother weigh in kilograms?

3.   Tim is painting his storage shed. He buys 4 gallons of white paint and 3 gallons of blue paint. Each gallon of white paint costs $15.72, and each gallon of blue paint is $21.87. How much will Tim spend in all on paint?

4.   Ribbon is sold at 3 yards for $6.33. Jackie bought 24 yards of ribbon for a project. How much did she pay?

Lesson 12:   Multiply a decimal fraction by single-digit whole numbers, including using estimation to confirm the placement of the decimal point.

Note: The use of unit language (e.g., 21 hundredths rather than 0.21) allows students to use knowledge of basic facts to compute easily with decimals.

1. Complete the sentence with the correct number of units, and then complete the equation.

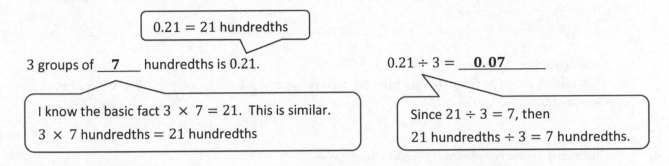

0.21 = 21 hundredths

3 groups of __7__ hundredths is 0.21.

0.21 ÷ 3 = __0.07__

I know the basic fact 3 × 7 = 21. This is similar.

3 × 7 hundredths = 21 hundredths

Since 21 ÷ 3 = 7, then

21 hundredths ÷ 3 = 7 hundredths.

2. Complete the number sentence. Express the quotient in units and then in standard form.

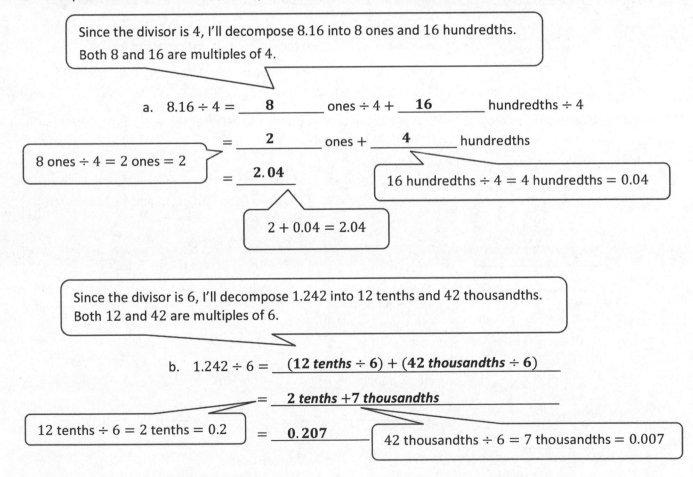

Since the divisor is 4, I'll decompose 8.16 into 8 ones and 16 hundredths. Both 8 and 16 are multiples of 4.

a. 8.16 ÷ 4 = __8__ ones ÷ 4 + __16__ hundredths ÷ 4

= __2__ ones + __4__ hundredths

8 ones ÷ 4 = 2 ones = 2

= __2.04__

16 hundredths ÷ 4 = 4 hundredths = 0.04

2 + 0.04 = 2.04

Since the divisor is 6, I'll decompose 1.242 into 12 tenths and 42 thousandths. Both 12 and 42 are multiples of 6.

b. 1.242 ÷ 6 = __(12 *tenths* ÷ 6) + (42 *thousandths* ÷ 6)__

= __2 *tenths* +7 *thousandths*__

12 tenths ÷ 6 = 2 tenths = 0.2

= __0.207__

42 thousandths ÷ 6 = 7 thousandths = 0.007

3. Find the quotients. Then, use words, numbers, or pictures to describe any relationships you notice between the pair of problems and their quotients.

   a. $35 \div 5 =$ ___**7**___   b. $3.5 \div 5 =$ ___**0.7**___

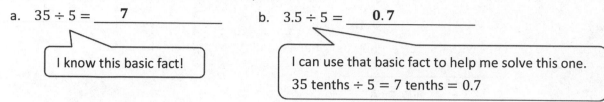

   I know this basic fact!

   I can use that basic fact to help me solve this one.
   35 tenths ÷ 5 = 7 tenths = 0.7

   Both problems are dividing by 5, but the quotient for part (a) is 10 times larger than the quotient for (b). That makes sense because the number we started with in part (a) is also 10 times larger than the number we started with in part (b).

4. Is the quotient below reasonable? Explain your answer.

   a. $0.56 \div 7 = 8$

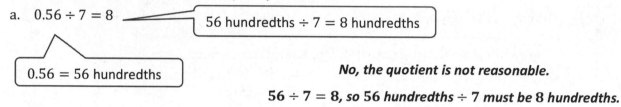

   56 hundredths ÷ 7 = 8 hundredths

   0.56 = 56 hundredths

   No, the quotient is not reasonable.

   $56 \div 7 = 8$, so 56 hundredths ÷ 7 must be 8 hundredths.

5. A toy airplane weighs 3.69 kg. It weighs 3 times as much as a toy car. What is the weight of the toy car?

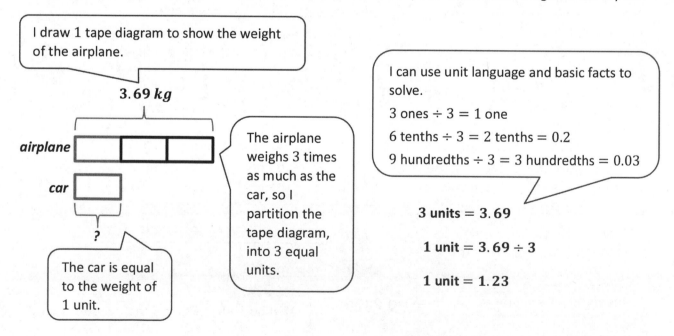

   I draw 1 tape diagram to show the weight of the airplane.

   3.69 kg

   airplane

   car

   ?

   The car is equal to the weight of 1 unit.

   The airplane weighs 3 times as much as the car, so I partition the tape diagram, into 3 equal units.

   I can use unit language and basic facts to solve.
   3 ones ÷ 3 = 1 one
   6 tenths ÷ 3 = 2 tenths = 0.2
   9 hundredths ÷ 3 = 3 hundredths = 0.03

   3 units = 3.69

   1 unit = 3.69 ÷ 3

   1 unit = 1.23

   The toy car weighs 1.23 kg.

Lesson 13:   Divide decimals by single-digit whole numbers involving easily identifiable multiples using place value understanding and relate to a written method.
© 2018 Great Minds®. eureka-math.org

Name _____   Date _____

1. Complete the sentences with the correct number of units, and then complete the equation.

    a. 3 groups of _____ tenths is 1.5. $1.5 \div 3 =$ _____

    b. 6 groups of _____ hundredths is 0.24. $0.24 \div 6 =$ _____

    c. 5 groups of _____ thousandths is 0.045. $0.045 \div 5 =$ _____

2. Complete the number sentence. Express the quotient in units and then in standard form.

    a. $9.36 \div 3 =$ _____ ones $\div$ 3 + _____ hundredths $\div$ 3

        = _____ ones + _____ hundredths

        = _____

    b. $36.012 \div 3 =$ _____ ones $\div$ 3 + _____ thousandths $\div$ 3

        = _____ ones + _____ thousandths

        = _____

    c. $3.55 \div 5 =$ _____ tenths $\div$ 5 + _____ hundredths $\div$ 5

        = _____

        = _____

    d. $3.545 \div 5 =$ _____

        = _____

        = _____

3. Find the quotients. Then, use words, numbers, or pictures to describe any relationships you notice between each pair of problems and quotients.

   a.   21 ÷ 7 = _____   2.1 ÷ 7 = _____

   b.   48 ÷ 8 = _____   0.048 ÷ 8 = _____

4. Are the quotients below reasonable? Explain your answers.

   a.   0.54 ÷ 6 = 9

   b.   5.4 ÷ 6 = 0.9

   c.   54 ÷ 6 = 0.09

Lesson 13:   Divide decimals by single-digit whole numbers involving easily identifiable multiples using place value understanding and relate to a written method.
© 2018 Great Minds®. eureka-math.org

5. A toy airplane costs $4.84. It costs 4 times as much as a toy car. What is the cost of the toy car?

6. Julian bought 3.9 liters of cranberry juice, and Jay bought 8.74 liters of apple juice. They mixed the two juices together and then poured them equally into 2 bottles. How many liters of juice are in each bottle?

Lesson 13: Divide decimals by single-digit whole numbers involving easily identifiable multiples using place value understanding and relate to a written method.

© 2018 Great Minds®. eureka-math.org

63

1. Draw place value disks on the place value chart to solve. Show each step using the standard algorithm.

$4.272 \div 3 = \underline{1.424}$

> 4.272 is divided into 3 equal groups. There is 1.424 in each group.

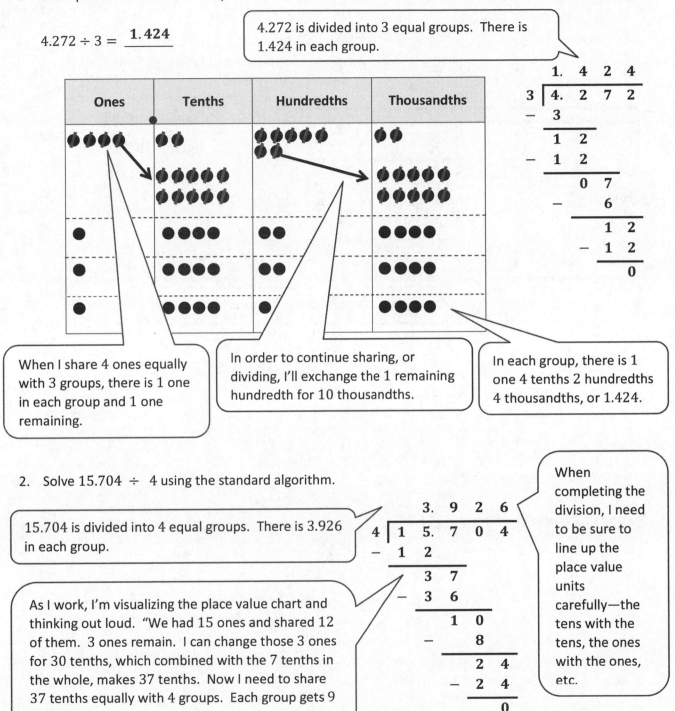

> When I share 4 ones equally with 3 groups, there is 1 one in each group and 1 one remaining.

> In order to continue sharing, or dividing, I'll exchange the 1 remaining hundredth for 10 thousandths.

> In each group, there is 1 one 4 tenths 2 hundredths 4 thousandths, or 1.424.

2. Solve $15.704 \div 4$ using the standard algorithm.

> 15.704 is divided into 4 equal groups. There is 3.926 in each group.

> As I work, I'm visualizing the place value chart and thinking out loud. "We had 15 ones and shared 12 of them. 3 ones remain. I can change those 3 ones for 30 tenths, which combined with the 7 tenths in the whole, makes 37 tenths. Now I need to share 37 tenths equally with 4 groups. Each group gets 9 tenths."

> When completing the division, I need to be sure to line up the place value units carefully—the tens with the tens, the ones with the ones, etc.

3.   Mr. Huynh paid $85.44 for 6 pounds of cashews. What's the cost of 1 pound of cashews?

> I'll draw a tape diagram and label it $85.44. Then I'll cut it equally into 6 units.

$85.44

6 units = $85.44

1 unit = $85.44 ÷ 6

        = $14.24

1 pound = ?

> To find the cost of 1 pound of cashews, I'll divide $85.44 by 6.

```
        1  4. 2  4
    6 | 8  5. 4  4
      - 6
        ‾‾‾‾
        2  5
      - 2  4
        ‾‾‾‾
           1  4
         - 1  2
           ‾‾‾‾
              2  4
            - 2  4
              ‾‾‾‾
                 0
```

> I'll solve using the long division algorithm.

The cost of 1 pound of cashews is $14.24.

Lesson 14:   Divide decimals with a remainder using place value understanding and relate to a written method.

© 2018 Great Minds®. eureka-math.org

Name _____   Date _____

1. Draw place value disks on the place value chart to solve. Show each step using the standard algorithm.

   a.   5.241 ÷ 3 = _____

Ones	Tenths	Hundredths	Thousandths

   $$3\overline{)5.241}$$

   b.   5.372 ÷ 4 = _____

Ones	Tenths	Hundredths	Thousandths

   $$4\overline{)5.372}$$

Lesson 14:   Divide decimals with a remainder using place value understanding and relate to a written method.

67

EUREKA
MATH®

© 2018 Great Minds®. eureka-math.org

2. Solve using the standard algorithm.

a. 0.64 ÷ 4 = _____	b. 6.45 ÷ 5 = _____	c. 16.404 ÷ 6 = _____

3. Mrs. Mayuko paid $40.68 for 3 kg of shrimp. What's the cost of 1 kilogram of shrimp?

4. The total weight of 6 pieces of butter and a bag of sugar is 3.8 lb. If the weight of the bag of sugar is 1.4 lb, what is the weight of each piece of butter?

Lesson 14:   Divide decimals with a remainder using place value understanding
and relate to a written method.

© 2018 Great Minds®. eureka-math.org

1. Draw place value disks on the place value chart to solve. Show each step in the standard algorithm.

$5.3 \div 4 = \underline{\textbf{1.325}}$

5.3 is divided into 4 equal groups
There is 1.325 in each group.

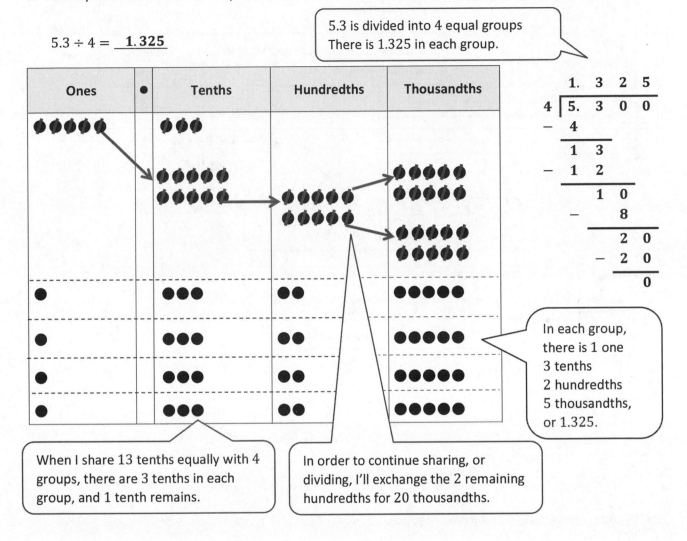

1. 3 2 5
4) 5. 3 0 0
− 4
1 3
− 1 2
1 0
− 8
2 0
− 2 0
0

In each group,
there is 1 one
3 tenths
2 hundredths
5 thousandths,
or 1.325.

When I share 13 tenths equally with 4 groups, there are 3 tenths in each group, and 1 tenth remains.

In order to continue sharing, or dividing, I'll exchange the 2 remaining hundredths for 20 thousandths.

2. Solve using the standard algorithm.

$9 \div 5 = \underline{\textbf{1.8}}$

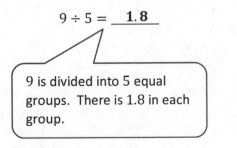

1. 8
5) 9. 0
− 5
4 0
− 4 0
0

9 is divided into 5 equal groups. There is 1.8 in each group.

In order to continue dividing, I'll rename the 4 remaining ones as 40 tenths.

40 tenths ÷ 5 = 8 tenths

 EUREKA MATH

Lesson 15: Divide decimals using place value understanding, including remainders in the smallest unit.

69

3. Four bakers shared 5.4 kilograms of sugar equally. How much sugar did they each receive?

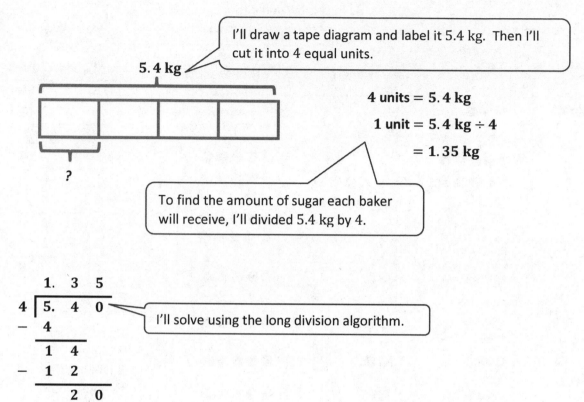

I'll draw a tape diagram and label it 5.4 kg. Then I'll cut it into 4 equal units.

5.4 kg

4 units = 5.4 kg

1 unit = 5.4 kg ÷ 4

= 1.35 kg

?

To find the amount of sugar each baker will receive, I'll divided 5.4 kg by 4.

$$
\begin{array}{r}
1.35 \\
4\overline{)5.40} \\
-4 \\
\hline
14 \\
-12 \\
\hline
20 \\
-20 \\
\hline
0
\end{array}
$$

I'll solve using the long division algorithm.

Each baker received 1.35 kilograms of sugar.

Lesson 15:   Divide decimals using place value understanding, including remainders in the smallest unit.

Name _____ Date _____

1. Draw place value disks on the place value chart to solve. Show each step in the standard algorithm.

a. 0.7 ÷ 4 = _____

Ones	•	Tenths	Hundredths	Thousandths

$$4\overline{)0.7}$$

b. 8.1 ÷ 5 = _____

Ones	•	Tenths	Hundredths	Thousandths

$$5\overline{)8.1}$$

EUREKA
MATH

Lesson 15:   Divide decimals using place value understanding, including remainders
             in the smallest unit.

© 2018 Great Minds®. eureka-math.org

71

2. Solve using the standard algorithm.

a. $0.7 \div 2 =$	b. $3.9 \div 6 =$	c. $9 \div 4 =$
d. $0.92 \div 2 =$	e. $9.4 \div 4 =$	f. $91 \div 8 =$

3. A rope 8.7 meters long is cut into 5 equal pieces. How long is each piece?

4. Yasmine bought 6 gallons of apple juice. After filling up 4 bottles of the same size with apple juice, she had 0.3 gallon of apple juice left. How many gallons of apple juice are in each container?

Lesson 15: Divide decimals using place value understanding, including remainders in the smallest unit.

1. A comic book costs $6.47, and a cookbook costs $9.79.

   a. Zion buys 5 comic books and 3 cookbooks. What is the total cost for all of the books?

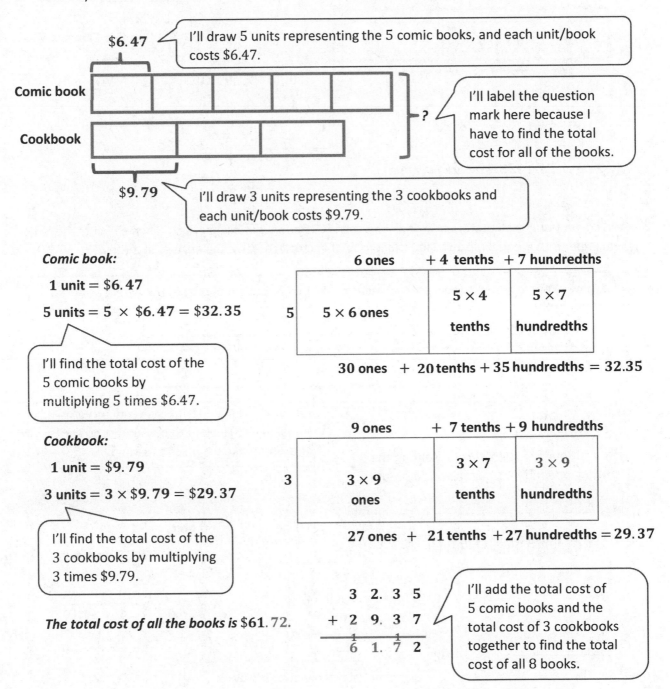

$6.47

I'll draw 5 units representing the 5 comic books, and each unit/book costs $6.47.

Comic book

Cookbook

?

I'll label the question mark here because I have to find the total cost for all of the books.

$9.79

I'll draw 3 units representing the 3 cookbooks and each unit/book costs $9.79.

Comic book:

1 unit = $6.47

5 units = 5 × $6.47 = $32.35

I'll find the total cost of the 5 comic books by multiplying 5 times $6.47.

	6 ones	+ 4 tenths	+ 7 hundredths
5	5 × 6 ones	5 × 4 tenths	5 × 7 hundredths

30 ones + 20 tenths + 35 hundredths = 32.35

Cookbook:

1 unit = $9.79

3 units = 3 × $9.79 = $29.37

I'll find the total cost of the 3 cookbooks by multiplying 3 times $9.79.

	9 ones	+ 7 tenths	+ 9 hundredths
3	3 × 9 ones	3 × 7 tenths	3 × 9 hundredths

27 ones + 21 tenths + 27 hundredths = 29.37

The total cost of all the books is $61.72.

```
  3 2. 3 5
+ 2 9. 3 7
─────────
  6 1. 7 2
```

I'll add the total cost of 5 comic books and the total cost of 3 cookbooks together to find the total cost of all 8 books.

EUREKA MATH

b.   Zion wants to pay for the all the books with a $100 bill. How much change will he get back?

$100

$61.72   ?

I'll subtract $61.72 from $100 to find Zion's change.

$100 − $61.72 = $38.28

$$\begin{array}{r} 0\ \ 9\ \ 9\ \ 9\ \ 10 \\ \cancel{1}\ \cancel{0}\ \cancel{0}.\ \cancel{0}\ \cancel{0} \\ -\ \ \ \ \ 6\ \ 1.\ 7\ \ 2 \\ \hline 3\ \ 8.\ 2\ \ 8 \end{array}$$

Zion will get $38.28 *back in change.*

2.   Ms. Porter bought 40 meters of string. She used 8.5 meters to tie a package. Then she cuts the remainder into 6 equal pieces. Find the length of each piece. Give the answer in meters.

I'll draw a tape diagram to represent the string Ms. Porter bought and label the whole as 40 m.

40 m

8.5 m   ?

I'll cut the remainder of the tape into 6 equal units. The length of 1 unit represents the length of each piece of string.

I'll cut out a small part representing the string needed for tying the package and label it 8.5 m.

6 units = 31.5 m

1 unit = 31.5 m ÷ 6 = 5.25 m

40 m − 8.5 m = 31.5 m

I can subtract 8.5 from 40 to find the length of the remaining string.

$$\begin{array}{r} 3\ \ 9\ \ 10 \\ \cancel{4}\ \cancel{0}.\ \cancel{0} \\ -\ \ \ \ 8.\ 5 \\ \hline 3\ \ 1.\ 5 \end{array}$$

I can divide 31.5 by 6 to find the length of each piece of string.

$$\begin{array}{r} 5.\ 2\ \ 5 \\ 6\,\overline{)\,3\ \ 1.\ 5\ \ 0} \\ -\ 3\ \ 0 \\ \hline 1\ \ 5 \\ -\ 1\ \ 2 \\ \hline 3\ \ 0 \\ \\ \hline 0 \end{array}$$

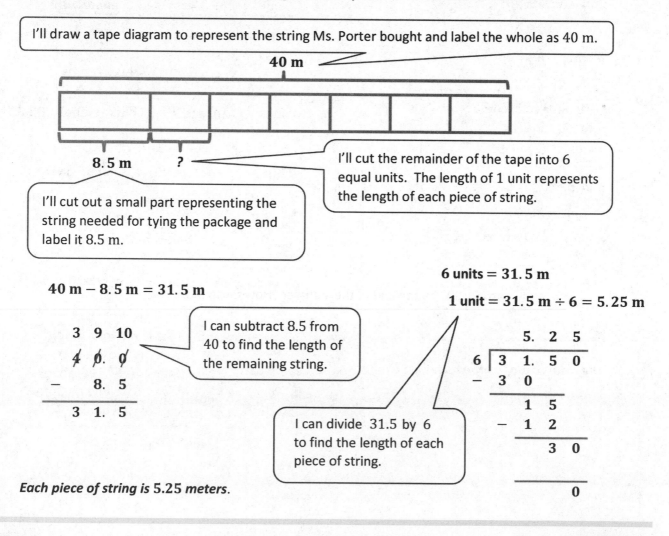

Each piece of string is 5.25 *meters.*

EUREKA MATH®

Name _____   Date _____

Solve using tape diagrams.

1. A gardener installed 42.6 meters of fencing in a week. He installed 13.45 meters on Monday and 9.5 meters on Tuesday. He installed the rest of the fence in equal lengths on Wednesday through Friday. How many meters of fencing did he install on each of the last three days?

2. Jenny charges $9.15 an hour to babysit toddlers and $7.45 an hour to babysit school-aged children.

   a. If Jenny babysat toddlers for 9 hours and school-aged children for 6 hours, how much money did she earn in all?

   b. Jenny wants to earn $1,300 by the end of the summer. How much more will she need to earn to meet her goal?

3. A table and 8 chairs weigh 235.68 lb together. If the table weighs 157.84 lb, what is the weight of one chair in pounds?

4. Mrs. Cleaver mixes 1.24 liters of red paint with 3 times as much blue paint to make purple paint. She pours the paint equally into 5 containers. How much blue paint is in each container? Give your answer in liters.

Lesson 16:   Solve word problems using decimal operations.

EUREKA
MATH

Grade 5
Module 2

1. Fill in the blanks using your knowledge of place value units and basic facts.

   a. 34×20

   Think: 34 ones × 2 tens = **68 tens**

   $34 \times 20 =$ **680**

   > 34 ones × 2 tens = $(34 \times 1) \times (2 \times 10)$.
   > First, I did the mental math: $34 \times 2 = 68$.
   > Then I thought about the units. *Ones times tens is tens*.
   > 68 tens is the same as 680 ones or 680.

   b. 420×20

   Think: 42 tens × 2 tens = **84 hundreds**

   $420 \times 20 =$ **8,400**

   > First, I'll multiply 42 times 2 in my head because that's a basic fact: 84.
   > Next, I have to think about the units. *Tens times tens is hundreds*.
   > Therefore, my answer is 84 hundreds or 8,400.

   > Another way to think about this is $42 \times 10 \times 2 \times 10$.
   > I can use the associative property to switch the order of the factors: $42 \times 2 \times 10 \times 10$.

   c. 400×500

   4 hundreds × 5 hundreds = **20 ten thousands**

   $400 \times 500 =$ **200,000**

   > I have to be careful because the basic fact, $4 \times 5 = 20$, ends in a zero.

   > Another way to think about this is $4 \times 100 \times 5 \times 100$
   > $= 4 \times 5 \times 100 \times 100$
   > $= 20 \times 100 \times 100$
   > $= 20 \times 10,000$
   > $= 200,000$

Lesson 1:   Multiply multi-digit whole numbers and multiples of 10 using place value patterns and the distributive and associative properties.

2. Determine if these equations are true or false. Defend your answer using knowledge of place value and the commutative, associate, and/or distributive properties.

a. 9 tens = 3 tens × 3 tens

   False. The basic fact is correct: $3 \times 3 = 9$.

   However, the units are not correct: 10×10 is 100.

   > Correct answers could be 9 tens = 3 tens × 3 ones, or 9 hundreds = 3 tens × 3 tens.

b. $93 \times 7 \times 100 = 930 \times 7 \times 10$

   True. I can rewrite the problem. $93 \times 7 \times (10 \times 10) = (93 \times 10) \times 7 \times 10$

   > The associative property tells me that I can group the factors in any order without changing the product.

3. Find the products. Show your thinking.

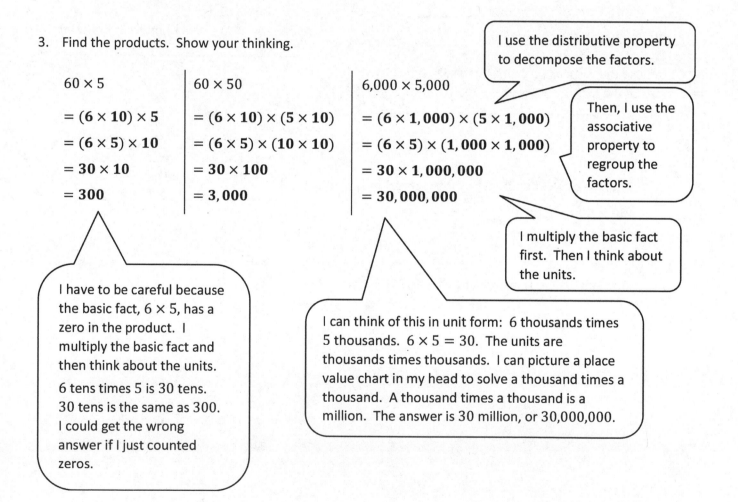

> I use the distributive property to decompose the factors.

60×5

$= (6 \times 10) \times 5$

$= (6 \times 5) \times 10$

$= 30 \times 10$

$= 300$

60×50

$= (6 \times 10) \times (5 \times 10)$

$= (6 \times 5) \times (10 \times 10)$

$= 30 \times 100$

$= 3,000$

$6,000 \times 5,000$

$= (6 \times 1,000) \times (5 \times 1,000)$

$= (6 \times 5) \times (1,000 \times 1,000)$

$= 30 \times 1,000,000$

$= 30,000,000$

> Then, I use the associative property to regroup the factors.

> I multiply the basic fact first. Then I think about the units.

> I have to be careful because the basic fact, 6×5, has a zero in the product. I multiply the basic fact and then think about the units.
>
> 6 tens times 5 is 30 tens. 30 tens is the same as 300. I could get the wrong answer if I just counted zeros.

> I can think of this in unit form: 6 thousands times 5 thousands. $6 \times 5 = 30$. The units are thousands times thousands. I can picture a place value chart in my head to solve a thousand times a thousand. A thousand times a thousand is a million. The answer is 30 million, or 30,000,000.

Lesson 1:   Multiply multi-digit whole numbers and multiples of 10 using place value patterns and the distributive and associative properties.

EUREKA MATH

Name _____   Date _____

1. Fill in the blanks using your knowledge of place value units and basic facts.

    a. 43 × 30

        Think: 43 ones × 3 tens = _____ tens

        43 × 30 = _____

    b. 430 × 30

        Think: 43 tens × 3 tens = _____ hundreds

        430 × 30 = _____

    c. 830 × 20

        Think: 83 tens × 2 tens = 166 _____

        830 × 20 = _____

    d. 4,400 × 400

        _____ hundreds × _____ hundreds = 176 _____

        4,400 × 400 = _____

    e. 80 × 5,000

        _____ tens × _____ thousands = 40 _____

        80 × 5,000 = _____

2. Determine if these equations are true or false. Defend your answer using your knowledge of place value and the commutative, associative, and/or distributive properties.

    a. 35 hundreds = 5 tens × 7 tens

    b. 770 × 6 = 77 × 6 × 100

    c. 50 tens × 4 hundreds = 40 tens × 5 hundreds

    d. 24 × 10 × 90 = 90 × 2,400

Lesson 1:   Multiply multi-digit whole numbers and multiples of 10 using place value patterns and the distributive and associative properties.

© 2018 Great Minds®. eureka-math.org

81

3. Find the products. Show your thinking. The first row gives some ideas for showing your thinking.

a. 5×5
    $= 25$

5×50
$= 25 \times 10$
$= 250$

50×50
$= (5 \times 10) \times (5 \times 10)$
$= (5 \times 5) \times 100$
$= 2,500$

50×500
$= (5 \times 5) \times (10 \times 100)$
$= 25,000$

b. 80×5 80×50 800×500 $8,000 \times 50$

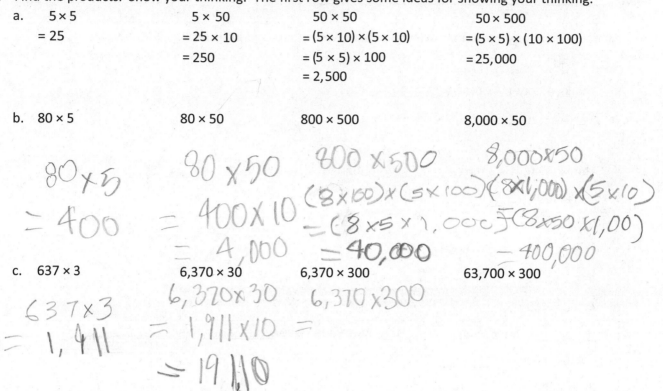

80×5
$= 400$

80×50
$= 400 \times 10$
$= 4,000$

800×500
$(8 \times 100) \times (5 \times 100)$
$= (8 \times 5 \times 1,000)$
$= 40,000$

$8,000 \times 50$
$(8 \times 1,000) \times (5 \times 10)$
$(8 \times 50 \times 1,00)$
$= 400,000$

c. 637×3 $6,370 \times 30$ $6,370 \times 300$ $63,700 \times 300$

637×3
$= 1,911$

$6,370 \times 30$
$= 1,911 \times 10$
$= 19,110$

$6,370 \times 300$
$=$

4. A concrete stepping-stone measures 20 square inches. What is the area of 30 such stones?

5. A number is 42,300 when multiplied by 10. Find the product of this number and 500.

Lesson 1:   Multiply multi-digit whole numbers and multiples of 10 using place value
patterns and the distributive and associative properties.

© 2018 Great Minds®. eureka-math.org

EUREKA
MATH

1. Round the factors to estimate the products.

I round each factor to the largest unit. For example, 387 rounds to 400.

The largest unit in 51 is tens. So, I round 51 to the nearest 10, which is 50.

a. $387 \times 51 \approx$ __**400**__ $\times$ __**50**__ $=$ __**20,000**__

Now that I have 2 rounded factors, I can use the distributive property to decompose the numbers. $400 \times 50 = (4 \times 100) \times (5 \times 10)$

I can use the associative property to regroup the factors.

$(4 \times 5) \times (100 \times 10) = 20 \times 1,000 = 20,000$

b. $6,286 \times 26 \approx$ __**6,000**__ $\times$ __**25**__ $=$ __**150,000**__

I could have chosen to round 25 to 30. However, multiplying by 25 is mental math for me. If I round 26 to 25, I know my estimated product will be closer to the actual product than if I round 26 to 30.

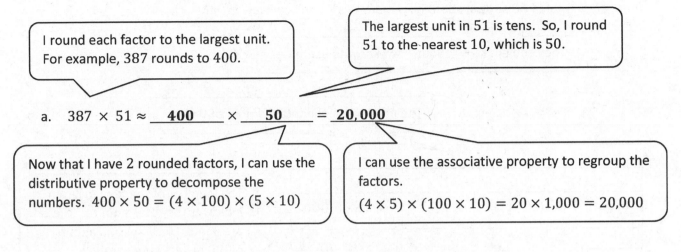

$$\begin{array}{r} {\scriptstyle 1}\ {\scriptstyle 2} \\ 637 \\ \times\quad 3 \\ \hline 1,911 \end{array}$$

EUREKA MATH

Lesson 2: Estimate multi-digit products by rounding factors to a basic fact and using place value patterns.

© 2018 Great Minds®. eureka-math.org

83

2. There are 6,015 seats available for each of the Radio City Rockettes Spring Spectacular dance shows. If there are a total of 68 shows, about how many tickets are available in all?

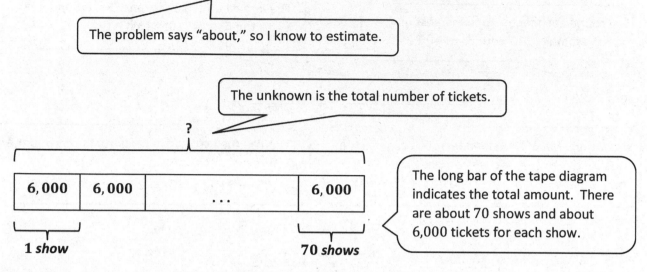

The problem says "about," so I know to estimate.

The unknown is the total number of tickets.

The long bar of the tape diagram indicates the total amount. There are about 70 shows and about 6,000 tickets for each show.

$6,000 \times 70$

$= 6 \text{ thousands} \times 7 \text{ tens} = 42 \text{ ten thousands} = 420,000$

$= (6 \times 7) \times (1,000 \times 10) = 42 \times 10,000 = 420,000$

About 420,000 tickets are available for the shows.

I can think about the problem in more than one way.

Lesson 2: Estimate multi-digit products by rounding factors to a basic fact and using place value patterns.

EUREKA MATH®

Name _____   Date _____

1. Round the factors to estimate the products.

    a. 697 × 82 ≈ _____ × _____ = _____

        A reasonable estimate for 697 × 82 is _____.

    b. 5,897 × 67 ≈ _____ × _____ = _____

        A reasonable estimate for 5,897 × 67 is _____.

    c. 8,840 × 45 ≈ _____ × _____ = _____

        A reasonable estimate for 8,840 × 45 is _____.

2. Complete the table using your understanding of place value and knowledge of rounding to estimate the product.

Expressions	Rounded Factors	Estimate
a. 3,409 × 73	3,000 × 70	210,000
b. 82,290 × 240		
c. 9,832 × 39		
d. 98 tens × 36 tens		
e. 893 hundreds × 85 tens		

3. The estimated answer to a multiplication problem is 800,000. Which of the following expressions could result in this answer? Explain how you know.

    8,146 × 12   81,467 × 121   8,146 × 121   81,477 × 1,217

Lesson 2:   Estimate multi-digit products by rounding factors to a basic fact and
             using place value patterns.

© 2018 Great Minds®. eureka-math.org

85

4. Fill in the blank with the missing estimate.

   a. $751 \times 34 \approx$ _____ $\times$ _____ = 24,000

   b. $627 \times 674 \approx$ _____ $\times$ _____ = 420,000

   c. $7,939 \times 541 \approx$ _____ $\times$ _____ = 4,000,000

5. In a single season, the New York Yankees sell an average of 42,362 tickets for each of their 81 home games. About how many tickets do they sell for an entire season of home games?

6. Raphael wants to buy a new car.

   a. He needs a down payment of $3,000. If he saves $340 each month, about how many months will it take him to save the down payment?

   b. His new car payment will be $288 each month for five years. Estimate the tot al of these payments.

Lesson 2: Estimate multi-digit products by rounding factors to a basic fact and using place value patterns.

© 2018 Great Minds®. eureka-math.org

1. Draw a model. Then write the numerical expression.

   a. The sum of 5 and 4, doubled

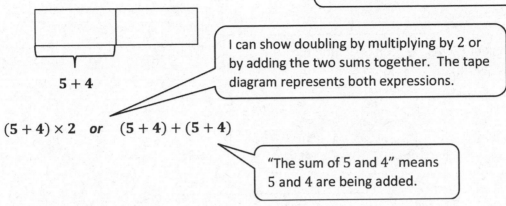

The directions don't ask me to solve, or evaluate, so I don't have to find the answers.

5 + 4

I can show doubling by multiplying by 2 or by adding the two sums together. The tape diagram represents both expressions.

$(5 + 4) \times 2$ *or* $(5 + 4) + (5 + 4)$

"The sum of 5 and 4" means 5 and 4 are being added.

   b. 3 times the difference between 42.6 and 23.9

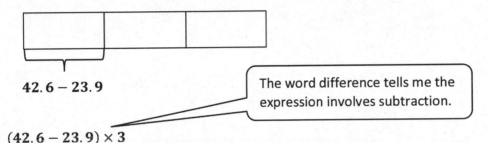

42. 6 − 23. 9

The word difference tells me the expression involves subtraction.

$(42. 6 − 23. 9) \times 3$

   c. The sum of 4 twelves and 3 sixes

Another way to say 4 *twelves* is to say 4 *groups of twelve*.

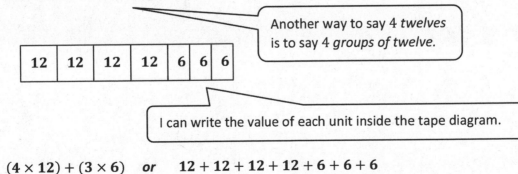

| 12 | 12 | 12 | 12 | 6 | 6 | 6 |

I can write the value of each unit inside the tape diagram.

$(4 \times 12) + (3 \times 6)$ *or* $12 + 12 + 12 + 12 + 6 + 6 + 6$

EUREKA MATH®

Lesson 3:   Write and interpret numerical expressions, and compare expressions using a visual model.

2. Compare the two expressions using >, <, or =.

a. $(2 \times 3) + (5 \times 3)$ $\boxed{=}$ $3 \times (2 + 5)$

Using the commutative property, I know that 7 threes is equal to 3 sevens.

I can think of $(2 \times 3) + (5 \times 3)$ in unit form.
2 threes + 5 threes = 7 threes = 21.

b. $28 \times (3 + 50)$ $\boxed{<}$ $(3 + 50) \times 82$

82 units of fifty-three is more than 28 units of fifty-three.

Lesson 3:   Write and interpret numerical expressions, and compare expressions
using a visual model.

Name _____ Date _____

1. Draw a model. Then, write the numerical expressions.

a. The sum of 21 and 4, doubled	b. 5 times the sum of 7 and 23
c. 2 times the difference between 49.5 and 37.5	d. The sum of 3 fifteens and 4 twos
e. The difference between 9 thirty-sevens and 8 thirty-sevens	f. Triple the sum of 45 and 55

Lesson 3: Write and interpret numerical expressions, and compare expressions
using a visual model.

© 2018 Great Minds®. eureka-math.org

89

2. Write the numerical expressions in words. Then, solve.

Expression	Words	The Value of the Expression
a. $10 \times (2.5 + 13.5)$		
b. $(98 - 78) \times 11$		
c. $(71 + 29) \times 26$		
d. $(50 \times 2) + (15 \times 2)$		

3. Compare the two expressions using > , < , or = . In the space beneath each pair of expressions, explain how you can compare without calculating. Draw a model if it helps you.

a. $93 \times (40 + 2)$	◯	$(40 + 2) \times 39$
b. 61×25	◯	60 twenty-fives minus 1 twenty-five

Lesson 3:   Write and interpret numerical expressions, and compare expressions using a visual model.

4. Larry claims that $(14 + 12) \times (8 + 12)$ and $(14 \times 12) + (8 \times 12)$ are equivalent because they have the same digits and the same operations.

a. Is Larry correct? Explain your thinking.

b. Which expression is greater? How much greater?

Lesson 3: Write and interpret numerical expressions, and compare expressions using a visual model.

91

1. Circle each expression that is not equivalent to the expression in **bold**.

   14 × 31

   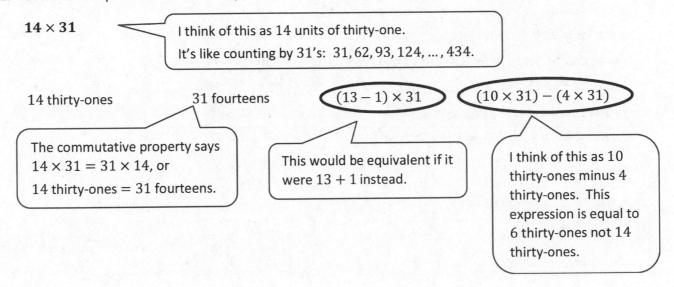

   I think of this as 14 units of thirty-one.
   It's like counting by 31's: 31, 62, 93, 124, ..., 434.

   14 thirty-ones   31 fourteens $(13 - 1) \times 31$ $(10 \times 31) - (4 \times 31)$

   The commutative property says
   $14 \times 31 = 31 \times 14$, or
   14 thirty-ones = 31 fourteens.

   This would be equivalent if it
   were $13 + 1$ instead.

   I think of this as 10
   thirty-ones minus 4
   thirty-ones. This
   expression is equal to
   6 thirty-ones not 14
   thirty-ones.

2. Solve using mental math. Draw a tape diagram and fill in the blanks to show your thinking.

   a. $19 \times 25 =$ __19__ twenty-fives

   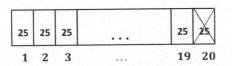

   Think: 20 twenty-fives − 1 twenty-five

   $= ($ __20__ $\times 25) - ($ __1__ $\times 25)$

   $=$ __500__ − __25__ $=$ __475__

   b. $21 \times 32 =$ __21__ thirty-twos

   | 32 | ... | 32 | 32 |

   20 *thirty-twos*

   Think: __20__ thirty-twos + __1__ thirty-two

   $= ($ __20__ $\times 32) + ($ __1__ $\times 32)$

   $=$ __640__ + __32__ $=$ __672__

3. The pet store has 99 fish tanks with 44 fish in each tank. How many fish does the pet store have? Use mental math to solve. Explain your thinking.

I need to find 99 forty-fours.

I know that 99 forty-fours is 1 unit of forty-four less than 100 forty-fours.

I multiplied 100 × 44, which is 4,400.

I need to subtract one group of 44.

4,400 − 44. The pet store has 4,356 fish.

Lesson 4: Convert numerical expressions into unit form as a mental strategy
for multi-digit multiplication.

Name _____ Date _____

1. Circle each expression that is not equivalent to the expression in **bold**.

   a. **37 × 19**

      37 nineteens   (30 × 19) – (7 × 29)   37 × (20 – 1)   (40 – 2) × 19

   b. **26 × 35**

      35 twenty-sixes   (26 + 30) × (26 + 5)   (26 × 30) + (26 × 5)   35 × (20 + 60)

   c. **34 × 89**

      34 × (80 + 9)   (34 × 8) + (34 × 9)   34 × (90 – 1)   89 thirty-fours

2. Solve using mental math. Draw a tape diagram and fill in the blanks to show your thinking. The first one is partially done for you.

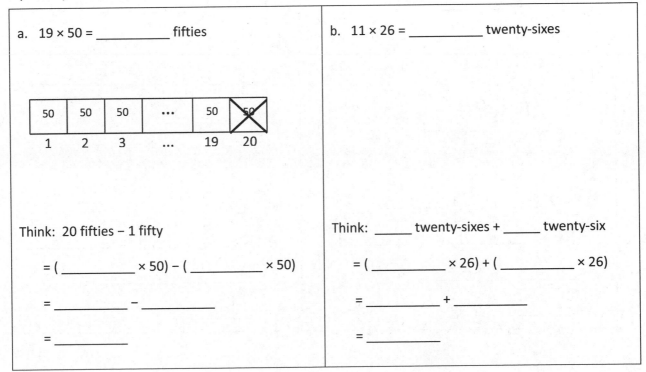

   a. 19 × 50 = _____ fifties

      | 50 | 50 | 50 | ... | 50 | 50 ✗ |
      | 1 | 2 | 3 | ... | 19 | 20 |

      Think: 20 fifties – 1 fifty

      = (_____ × 50) – (_____ × 50)

      = _____ – _____

      = _____

   b. 11 × 26 = _____ twenty-sixes

      Think: _____ twenty-sixes + _____ twenty-six

      = (_____ × 26) + (_____ × 26)

      = _____ + _____

      = _____

Lesson 4:   Convert numerical expressions into unit form as a mental strategy
             for multi-digit multiplication.

95

© 2018 Great Minds®. eureka-math.org

c. 49 × 12 = _____ twelves

Think: _____ twelves – 1 twelve

= (_____ × 12) – (_____ × 12)

= _____ – _____

= _____

d. 12 × 25 = _____ twenty-fives

Think: _____ twenty-fives + _____ twenty-fives

= (_____ × 25) + (_____ × 25)

= _____ + _____

= _____

3. Define the unit in word form and complete the sequence of problems as was done in the lesson.

a. 29 × 12 = 29 _____

Think: 30 _____ – 1 _____

= (30 × _____) – (1 × _____)

= _____ – _____

= _____

b. 11 × 31 = 31 _____

Think: 30 _____ + 1 _____

= (30 × _____) + (1 × _____)

= _____ + _____

= _____

Lesson 4:   Convert numerical expressions into unit form as a mental strategy
                for multi-digit multiplication.

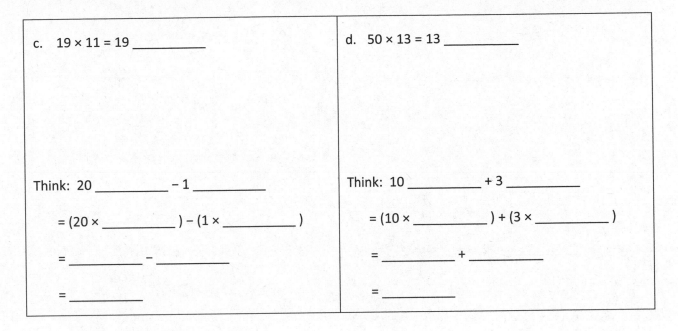

c. 19 × 11 = 19 _____

Think: 20 _____ − 1 _____

= (20 × _____) − (1 × _____)

= _____ − _____

= _____

d. 50 × 13 = 13 _____

Think: 10 _____ + 3 _____

= (10 × _____) + (3 × _____)

= _____ + _____

= _____

4. How can 12 × 50 help you find 12 × 49?

5. Solve mentally.

a. 16 × 99 = _____

b. 20 × 101 = _____

6. Joy is helping her father to build a rectangular deck that measures 14 ft by 19 ft. Find the area of the deck using a mental strategy. Explain your thinking.

7. The Lason School turns 101 years old in June. In order to celebrate, they ask each of the 23 classes to collect 101 items and make a collage. How many total items will be in the collage? Use mental math to solve. Explain your thinking.

Lesson 4:   Convert numerical expressions into unit form as a mental strategy for multi-digit multiplication.

97

© 2018 Great Minds®. eureka-math.org

1. Draw an area model, and then solve using the standard algorithm. Use arrows to match the partial products from the area model to the partial products in the algorithm.

a. 33 × 21

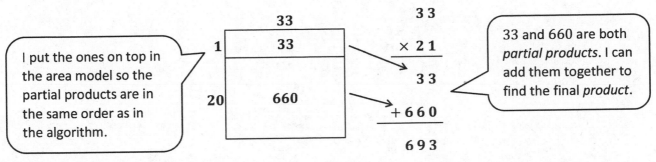

I put the ones on top in the area model so the partial products are in the same order as in the algorithm.

```
       33
      × 2 1
      ─────
        3 3
      +6 6 0
      ─────
        6 9 3
```

33 and 660 are both *partial products*. I can add them together to find the final *product*.

b. 433 × 21

	433
1	433
20	8,660

```
      4 3 3
    ×   2 1
    ───────
      4 3 3
   + 8,6 6 0
     1
    ───────
    9,0 9 3
```

When I add the hundreds in the two partial products, the sum is 10 hundreds, or 1,000. I record the 1 thousand below the partial products, rather than above.

2. Elizabeth pays $123 each month for her cell phone service. How much does she spend in a year?

I can draw an area model to help me see where the 2 partial products come from.

	123
2	246
10	1,230

```
       1 2 3
     ×   1 2
     ───────
       2 4 6
    + 1,2 3 0
     ───────
     1,4 7 6
```

Elizabeth spends $1,476 in a year for cell phone service.

 EUREKA MATH®

Lesson 5: Connect visual models and the distributive property to partial products of the standard algorithm without renaming.

© 2018 Great Minds®. eureka-math.org

99

Name __Adrienne__

Date _____

1. Draw an area model, and then solve using the standard algorithm. Use arrows to match the partial products from the area model to the partial products in the algorithm.

   a. $24 \times 21 =$ ___106___

   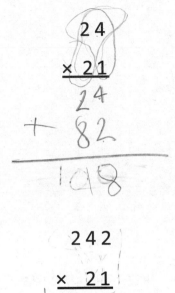

   ```
       2 4
     × 2 1
       2 4
   +   8 2
   ─────────
     1 0 8
   ```

   b. $242 \times 21 =$ ___726___

   ```
        2 4 2
      ×   2 1
       ⁺¹2 4 2
     ×   4 8 4
   ───────────
       7 2 8
   ```

2. Solve using the standard algorithm.

   a. $314 \times 22 =$ _____

   b. $413 \times 22 =$ _____

   c. $213 \times 32 =$ _____

   ```
       3 14
     ×  2 2
     ─────────
     ×  6 28
     6,2 80
   ─────────
     6,9 08
   ```

   ```
        413
      ×  22
   ─────────
     ⁺¹8 26
     8,2 60
   ─────────
     9,086
   ```

   ```
        213
      ×  32
   ─────────
     ⁺¹4 26
     6,3 40
   ─────────
     6,8 16
   ```

EUREKA MATH

Lesson 5: Connect visual models and the distributive property to partial products of the standard algorithm without renaming.

© 2018 Great Minds®. eureka-math.org

101

3. A young snake measures 0.23 meters long. During the course of his lifetime, he will grow to be 13 times his current length. What will his length be when he is full grown?

4. Zenin earns $142 per shift at his new job. During a pay period, he works 12 shifts. What would his pay be for that period?

Lesson 5: Connect visual models and the distributive property to partial products of the standard algorithm without renaming.

EUREKA
MATH

1. Draw an area model. Then, solve using the standard algorithm. Use arrows to match the partial products from your area model to the partial products in the algorithm.

   a. 39×45

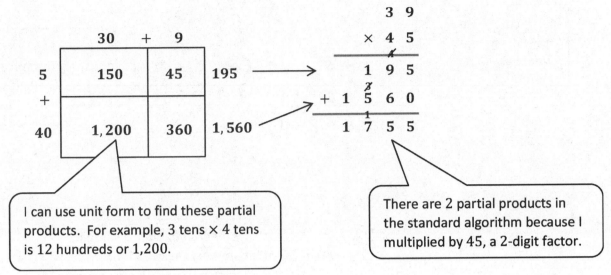

   I can use unit form to find these partial products. For example, 3 tens × 4 tens is 12 hundreds or 1,200.

   There are 2 partial products in the standard algorithm because I multiplied by 45, a 2-digit factor.

   b. 339×45

   The area model shows the factors expanded. If I wanted to, I could put the + between the units.

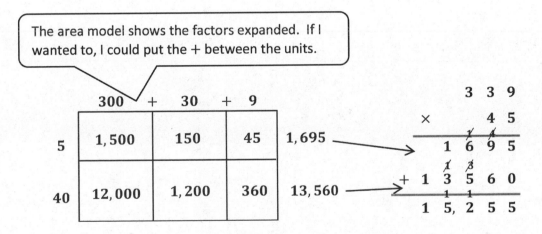

Lesson 6:   Connect area diagrams and the distributive property to partial products of the standard algorithm without renaming.

© 2018 Great Minds®. eureka-math.org

103

2. Desmond bought a car and paid monthly installments. Each installment was $452 per month. After 36 months, Desmond still owes $1,567. What was the total price of the car?

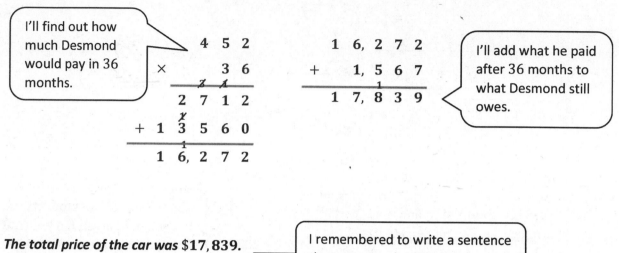

I'll find out how much Desmond would pay in 36 months.

```
      4 5 2
  ×     3 6
    2 7 1 2
+ 1 3 5 6 0
  1 6, 2 7 2
```

```
  1 6, 2 7 2
+    1, 5 6 7
  1 7, 8 3 9
```

I'll add what he paid after 36 months to what Desmond still owes.

The total price of the car was $17,839.

I remembered to write a sentence that answers the question.

Lesson 6: Connect area diagrams and the distributive property to partial products of the standard algorithm without renaming.

Name _____ Date _____

1. Draw an area model. Then, solve using the standard algorithm. Use arrows to match the partial products from your area model to the partial products in the algorithm.

   a. 27 × 36

   27
   × 36

   27
   × 36
   ____
   162
   6210
   ____
   6,372

   27
   ×

   b. 527 × 36

   527
   × 36
   ____
   722

   527
   × 36

   3162
   156,210
   ____
   159,372

EUREKA MATH

Lesson 6: Connect area diagrams and the distributive property to partial products of the standard algorithm without renaming.

105

© 2018 Great Minds®. eureka-math.org

2. Solve using the standard algorithm.

a. 649 × 53

$$
\begin{array}{r}
649 \\
\times\ 53 \\
\hline
1947 \\
+\ 50 \\
\end{array}
$$

b. 496 × 53

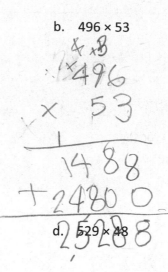

$$
\begin{array}{r}
496 \\
\times\ 53 \\
\hline
1488 \\
+\ 24800 \\
\hline
25,288 \\
\end{array}
$$

25,288

c. 758 × 46

$$
\begin{array}{r}
758 \\
\times\ 46 \\
\hline
\end{array}
$$

d. 529 × 48

3. Each of the 25 students in Mr. McDonald's class sold 16 raffle tickets. If each ticket costs $15, how much money did Mr. McDonald's students raise?

25   16 15

$$
\begin{array}{r}
25 \\
\times\ 16 \\
\hline
150 \\
+\ 250 \\
\hline
300 \\
\end{array}
$$

4. Jayson buys a car and pays by installments. Each installment is $567 per month. After 48 months, Jayson owes $1,250. What was the total price of the vehicle?

Lesson 6:   Connect area diagrams and the distributive property to partial products of the standard algorithm without renaming.

© 2018 Great Minds®. eureka-math.org

1. Draw an area model. Then, solve using the standard algorithm. Use arrows to match the partial products from the area model to the partial products in the algorithm.

$431 \times 246 = \underline{\textbf{106, 026}}$

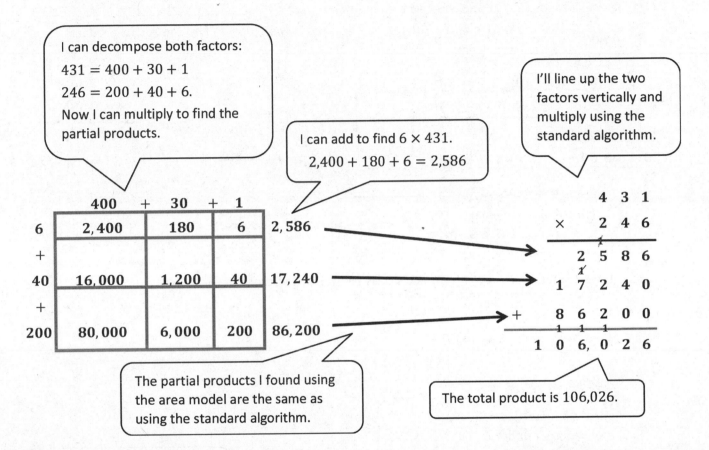

I can decompose both factors:
$431 = 400 + 30 + 1$
$246 = 200 + 40 + 6$.
Now I can multiply to find the partial products.

I'll line up the two factors vertically and multiply using the standard algorithm.

I can add to find 6×431.
$2,400 + 180 + 6 = 2,586$

The partial products I found using the area model are the same as using the standard algorithm.

The total product is 106,026.

Lesson 7: Connect area models and the distributive property to partial products of the standard algorithm with renaming.

© 2018 Great Minds®. eureka-math.org

107

2. Solve by drawing the area model and using the standard algorithm.

$2,451 \times 107 = \underline{\textbf{262, 257}}$

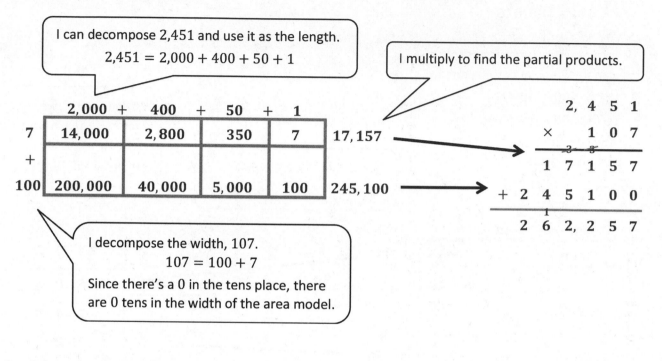

I can decompose 2,451 and use it as the length.
$2,451 = 2,000 + 400 + 50 + 1$

I multiply to find the partial products.

	2,000 +	400 +	50 +	1	
7	14,000	2,800	350	7	17,157
+					
100	200,000	40,000	5,000	100	245,100

$$\begin{array}{r} 2,\ 4\ 5\ 1 \\ \times\ \ \ \ 1\ 0\ 7 \\ \hline 1\ 7\ 1\ 5\ 7 \\ +\ 2\ 4\ 5\ 1\ 0\ 0 \\ \hline 2\ 6\ 2,\ 2\ 5\ 7 \end{array}$$

I decompose the width, 107.
$107 = 100 + 7$
Since there's a 0 in the tens place, there are 0 tens in the width of the area model.

3. Solve using the standard algorithm.

$7,302 \times 408 = \underline{\textbf{2, 979, 216}}$

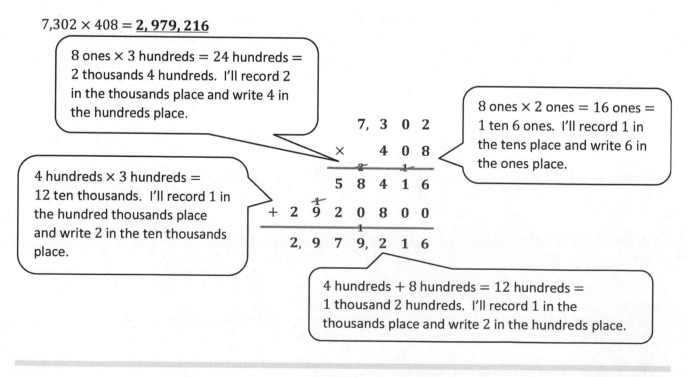

8 ones × 3 hundreds = 24 hundreds = 2 thousands 4 hundreds. I'll record 2 in the thousands place and write 4 in the hundreds place.

8 ones × 2 ones = 16 ones = 1 ten 6 ones. I'll record 1 in the tens place and write 6 in the ones place.

4 hundreds × 3 hundreds = 12 ten thousands. I'll record 1 in the hundred thousands place and write 2 in the ten thousands place.

$$\begin{array}{r} 7,\ 3\ 0\ 2 \\ \times\ \ \ \ 4\ 0\ 8 \\ \hline 5\ 8\ 4\ 1\ 6 \\ +\ 2\ 9\ 2\ 0\ 8\ 0\ 0 \\ \hline 2,\ 9\ 7\ 9,\ 2\ 1\ 6 \end{array}$$

4 hundreds + 8 hundreds = 12 hundreds = 1 thousand 2 hundreds. I'll record 1 in the thousands place and write 2 in the hundreds place.

Lesson 7: Connect area models and the distributive property to partial products of the standard algorithm with renaming.

EUREKA MATH®

Name _Adrienne_____ Date _____

1. Draw an area model. Then, solve using the standard algorithm. Use arrows to match the partial products from your area model to the partial products in your algorithm.

   a. 273 × 346

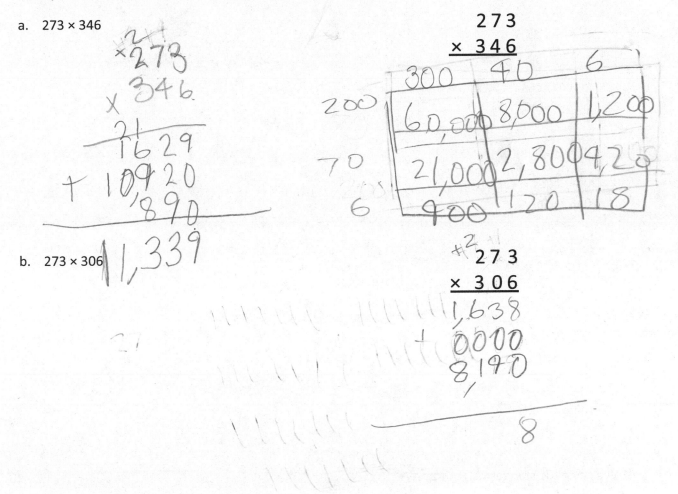

   b. 273 × 306

   c. Both Parts (a) and (b) have three-digit multipliers. Why are there three partial products in Part (a) and only two partial products in Part (b)?

Lesson 7:   Connect area models and the distributive property to partial products of the standard algorithm with renaming.

109

2. Solve by drawing the area model and using the standard algorithm.

   a.   7,481 × 290

   b.   7,018 × 209

3. Solve using the standard algorithm.
   a.   426 × 357

   b.   1,426 × 357

Lesson 7:   Connect area models and the distributive property to partial products
                              of the standard algorithm with renaming.

                    © 2018 Great Minds®. eureka-math.org

c. 426 × 307

d. 1,426 × 307

4. The Hudson Valley Renegades Stadium holds a maximum of 4,505 people. During the height of their popularity, they sold out 219 consecutive games. How many tickets were sold during this time?

5. One Saturday at the farmer's market, each of the 94 vendors made $502 in profit. How much profit did all vendors make that Saturday?

Lesson 7:   Connect area models and the distributive property to partial products of the standard algorithm with renaming.

© 2018 Great Minds®. eureka-math.org

111

1. Estimate the products first. Solve by using the standard algorithm. Use your estimate to check the reasonableness of the product.

   a. 795×248

   $\approx 800 \times 200$

   $= 160,000$

   > I could have rounded 248 to 250 in order to have an estimate that is closer to the actual product. Another reasonable estimate is $800 \times 250 = 200,000$.

   ```
         7  9  5
      ×  2  4  8
      _____
      6  3  6  0
   ```

   > $8 \times 5 = 40$, which I record as 4 tens 0 ones. 8×9 tens $= 72$ tens plus 4 tens, makes 76 tens. I record 76 tens as 7 hundreds 6 tens.

   ```
       3  1  8  0  0
   +   1  5  9  0  0  0
   _____
       1  9  7,  1  6  0
   ```

   > This product is reasonable because $197,160$ is close to $160,000$. My other estimate is also reasonable because $197,000$ is very close to $200,000$.

   b. $4,308 \times 505$

   $\approx 4,000 \times 500$

   $= 2,000,000$

   > I have to be careful to estimate accurately. 4 thousands × 5 hundreds is 20 hundred thousands. That's the same as 2 million. If I just count zeros I might get a wrong estimate.

   ```
       4,  3  0  8
   ×      5  0  5
   _____
       2  1  5  4  0
   ```

   > This partial product is the result of $5 \times 4,308$.

   ```
   +  2  1  5  4  0  0  0
   _____
      2,  1  7  5,  5  4  0
   ```

   > This partial product is the result of $500 \times 4,308$. It makes sense that it is 100 times greater than the first partial product.

2. When multiplying 809 times 528, Isaac got a product of 42,715. Without calculating, does his product seem reasonable? Explain your thinking.

   Isaac's product of about 40 thousands is not reasonable. A correct estimate is 8 hundreds times 5 hundreds, which is 40 ten thousands. That's the same as 400,000 not 40,000.

   > I think Isaac rounded 809 to 800 and 528 to 500. Then, I think he multiplied 8 times 5 to get 40. From there, I think he miscounted the zeros.

EUREKA MATH

Lesson 8: Fluently multiply multi-digit whole numbers using the standard algorithm and using estimation to check for reasonableness of the product.

© 2018 Great Minds®. eureka-math.org

113

Name _____   Date _____

1. Estimate the product first. Solve by using the standard algorithm. Use your estimate to check the reasonableness of the product.

a. 312 × 149	b. 743 × 295	c. 428 × 637
≈ 300 × 100 = 30,000 3 1 2 × 1 4 9 ‾‾‾‾‾‾‾		
d. 691 × 305	e. 4,208 × 606	f. 3,068 × 523
g. 430 × 3,064	h. 3,007 × 502	i. 254 × 6,104

Lesson 8:   Fluently multiply multi-digit whole numbers using the standard algorithm
and using estimation to check for reasonableness of the product.

115

© 2018 Great Minds®. eureka-math.org

2. When multiplying 1,729 times 308, Clayton got a product of 53,253. Without calculating, does his product seem reasonable? Explain your thinking.

3. A publisher prints 1,912 copies of a book in each print run. If they print 305 runs, the manager wants to know about how many books will be printed. What is a reasonable estimate?

Lesson 8:   Fluently multiply multi-digit whole numbers using the standard algorithm and using estimation to check for reasonableness of the product.

EUREKA MATH

Solve.

1. Howard and Robin are both cabinet makers. Over the last year, Howard made 107 cabinets. Robin made 28 more cabinets than Howard. Each cabinet they make has exactly 102 nails in it. How many nails did they use altogether while making the cabinets?

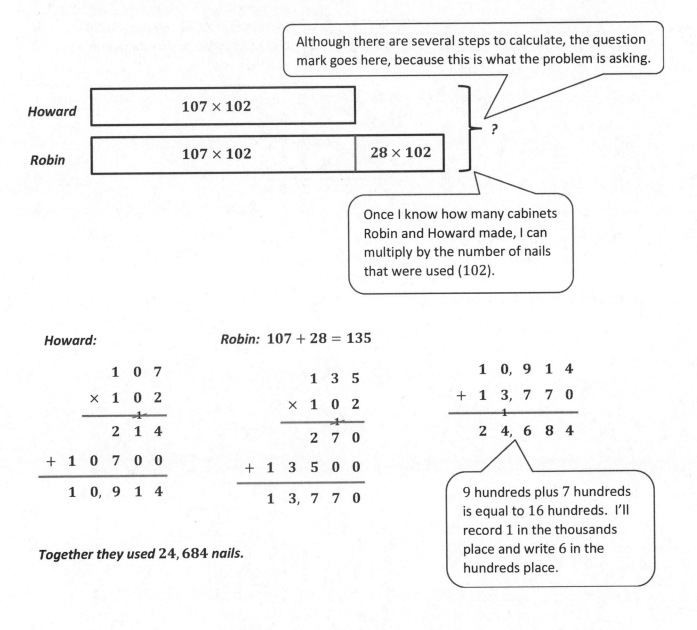

Although there are several steps to calculate, the question mark goes here, because this is what the problem is asking.

Howard 107×102

Robin 107×102 | 28×102

?

Once I know how many cabinets Robin and Howard made, I can multiply by the number of nails that were used (102).

Howard:

$$
\begin{array}{r}
1\ 0\ 7 \\
\times\ 1\ 0\ 2 \\
\hline
2\ 1\ 4 \\
+\ 1\ 0\ 7\ 0\ 0 \\
\hline
1\ 0,\ 9\ 1\ 4
\end{array}
$$

Robin: $107 + 28 = 135$

$$
\begin{array}{r}
1\ 3\ 5 \\
\times\ 1\ 0\ 2 \\
\hline
2\ 7\ 0 \\
+\ 1\ 3\ 5\ 0\ 0 \\
\hline
1\ 3,\ 7\ 7\ 0
\end{array}
$$

$$
\begin{array}{r}
1\ 0,\ 9\ 1\ 4 \\
+\ 1\ 3,\ 7\ 7\ 0 \\
\hline
2\ 4,\ 6\ 8\ 4
\end{array}
$$

9 hundreds plus 7 hundreds is equal to 16 hundreds. I'll record 1 in the thousands place and write 6 in the hundreds place.

Together they used 24, 684 nails.

Lesson 9:   Fluently multiply multi-digit whole numbers using the standard algorithm to solve multi-step word problems.

117

© 2018 Great Minds®. eureka-math.org

2. Mrs. Peterson made 32 car payments at $533 each. She still owes $8,530 on her car. How much did the car cost?

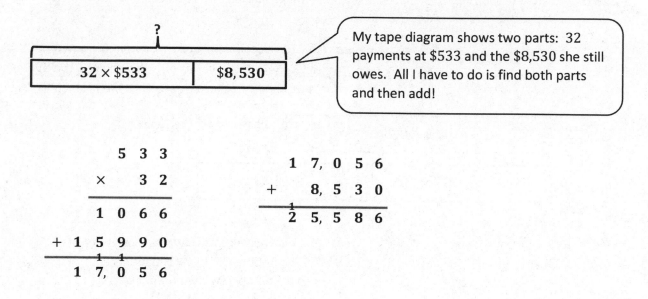

My tape diagram shows two parts: 32 payments at $533 and the $8,530 she still owes. All I have to do is find both parts and then add!

$$
\begin{array}{r}
5\ 3\ 3 \\
\times\quad 3\ 2 \\
\hline
1\ 0\ 6\ 6 \\
+\ 1\ 5\ 9\ 9\ 0 \\
\hline
1\ 7,\ 0\ 5\ 6
\end{array}
$$

$$
\begin{array}{r}
1\ 7,\ 0\ 5\ 6 \\
+\quad 8,\ 5\ 3\ 0 \\
\hline
2\ 5,\ 5\ 8\ 6
\end{array}
$$

Mrs. Peterson's car cost $25,586.

Lesson 9: Fluently multiply multi-digit whole numbers using the standard algorithm to solve multi-step word problems.

© 2018 Great Minds®. eureka-math.org

Name _____ Date _____

Solve.

1. Jeffery bought 203 sheets of stickers. Each sheet has a dozen stickers. He gave away 907 stickers to his family and friends on Valentine's Day. How many stickers does Jeffery have remaining?

2. During the 2011 season, a quarterback passed for 302 yards per game. He played in all 16 regular season games that year.

    a. For how many total yards did the quarterback pass?

    b. If he matches this passing total for each of the next 13 seasons, how many yards will he pass for in his career?

EUREKA
MATH

Lesson 9: Fluently multiply multi-digit whole numbers using the standard algorithm to solve multi-step word problems.

119

© 2018 Great Minds®. eureka-math.org

3. Bao saved $179 a month. He saved $145 less than Ada each month. How much would Ada save in three and a half years?

4. Mrs. Williams is knitting a blanket for her newborn granddaughter. The blanket is 2.25 meters long and 1.8 meters wide. What is the area of the blanket? Write the answer in centimeters.

Lesson 9: Fluently multiply multi-digit whole numbers using the standard algorithm to solve multi-step word problems.

5. Use the chart to solve.

Soccer Field Dimensions

	FIFA Regulation (in yards)	New York State High Schools (in yards)
Minimum Length	110	100
Maximum Length	120	120
Minimum Width	70	55
Maximum Width	80	80

a. Write an expression to find the difference in the maximum area and minimum area of a NYS high school soccer field. Then, evaluate your expression.

b. Would a field with a width of 75 yards and an area of 7,500 square yards be within FIFA regulation? Why or why not?

c. It costs $26 to fertilize, water, mow, and maintain each square yard of a full size FIFA field (with maximum dimensions) before each game. How much will it cost to prepare the field for next week's match?

Lesson 9:   Fluently multiply multi-digit whole numbers using the standard algorithm to solve multi-step word problems.

121

© 2018 Great Minds®. eureka-math.org

1. Estimate the product. Solve using an area model and the standard algorithm. Remember to express your products in standard form.

I rename 4.1 as 41 tenths and then multiply.

I round 23 to the nearest ten, 2 tens, and 4.1 to the nearest one, 4 ones.

$23 \times 4.1 \approx \underline{\textbf{20}} \times \underline{\textbf{4}} = \underline{\textbf{80}}$

2 tens × 4 ones = 8 tens, or 80. This is the estimated product.

$$\begin{array}{r} 2\ 3 \\ \times\ 4\ 1\ \text{(tenths)} \\ \hline 2\ 3 \\ +\ 9\ 2\ 0 \\ \hline 9\ 4\ 3\ \textit{(tenths)} = 94.3 \end{array}$$

943 tenths, or 94.3, is the actual product, which is close to my estimated product of 80.

I decompose 23 to 20 + 3, and 41 tenths to 40 tenths + 1 tenth.

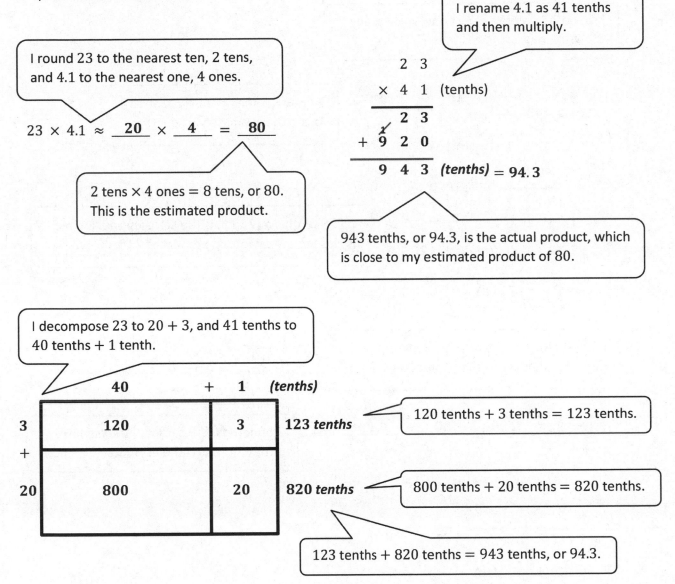

	40	+	1	(tenths)
3	120		3	123 *tenths*
+ 20	800		20	820 *tenths*

120 tenths + 3 tenths = 123 tenths.

800 tenths + 20 tenths = 820 tenths.

123 tenths + 820 tenths = 943 tenths, or 94.3.

Lesson 10: Multiply decimal fractions with tenths by multi-digit whole numbers using place value understanding to record partial products.

123

2. Estimate. Then, use the standard algorithm to solve. Express your products in standard form.

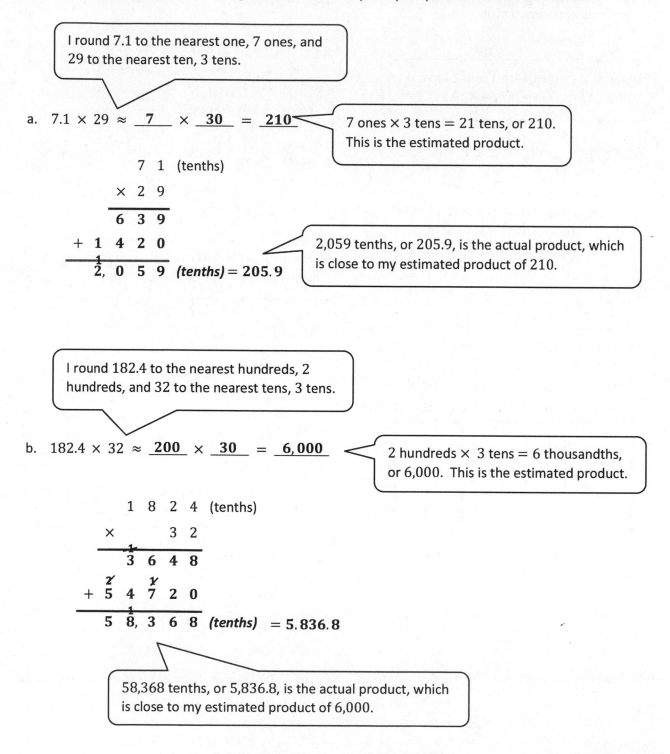

I round 7.1 to the nearest one, 7 ones, and 29 to the nearest ten, 3 tens.

a. $7.1 \times 29 \approx$ __7__ $\times$ __30__ = __210__

7 ones × 3 tens = 21 tens, or 210. This is the estimated product.

```
      7  1  (tenths)
   ×  2  9
   ─────────
      6  3  9
+  1  4  2  0
   ─────────
   2, 0  5  9  (tenths) = 205.9
```

2,059 tenths, or 205.9, is the actual product, which is close to my estimated product of 210.

I round 182.4 to the nearest hundreds, 2 hundreds, and 32 to the nearest tens, 3 tens.

b. $182.4 \times 32 \approx$ __200__ $\times$ __30__ = __6,000__

2 hundreds × 3 tens = 6 thousandths, or 6,000. This is the estimated product.

```
      1  8  2  4  (tenths)
   ×        3  2
   ──────────────
      3  6  4  8
+  5  4  7  2  0
   ──────────────
   5  8, 3  6  8  (tenths)  = 5,836.8
```

58,368 tenths, or 5,836.8, is the actual product, which is close to my estimated product of 6,000.

Lesson 10:   Multiply decimal fractions with tenths by multi-digit whole numbers using place value understanding to record partial products.

© 2018 Great Minds®. eureka-math.org

EUREKA MATH

Name ___Adrienne___   Date _____

1. Estimate the product. Solve using an area model and the standard algorithm. Remember to express your products in standard form.

   a. $53 \times 1.2 \approx$ ___50___ $\times$ ___1.0___ $=$ ___50___

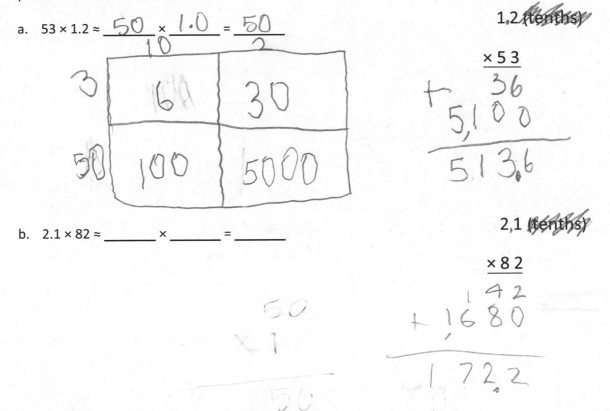

   1.2 (tenths)

   $\times 53$

   $+\ 36$

   $5,100$

   _____

   $5,13.6$

   b. $2.1 \times 82 \approx$ _____ $\times$ _____ $=$ _____

   2,1 (tenths)

   $\times 82$

   $\ \ 42$

   $+\ 1680$

   _____

   $1,72.2$

2. Estimate. Then, use the standard algorithm to solve. Express your products in standard form.

   a. $4.2 \times 34 \approx$ ___4.0___ $\times$ ___30___ $=$ ___1200___

   b. $65 \times 5.8 \approx$ _____ $\times$ _____ $=$ _____

   4 2 (tenths)

   $\times 34$

   $+\ 168$

   $12 60$

   _____

   $1 4 2.8$

   5 8 (tenths)

   $\times 65$

Lesson 10: Multiply decimal fractions with tenths by multi-digit whole numbers using place value understanding to record partial products.

125

c. 3.3 × 16 ≈ _____ × _____ = _____

d. 15.6 × 17 ≈ _____ × _____ = _____

e. 73 × 2.4 ≈ _____ × _____ = _____

f. 193.5 × 57 ≈ _____ × _____ = _____

3. Mr. Jansen is building an ice rink in his backyard that will measure 8.4 meters by 22 meters. What is the area of the rink?

4. Rachel runs 3.2 miles each weekday and 1.5 miles each day of the weekend. How many miles will she have run in 6 weeks?

Lesson 10: Multiply decimal fractions with tenths by multi-digit whole numbers using place value understanding to record partial products.

1. Estimate the product. Solve using the standard algorithm. Use the thought bubbles to show your thinking.

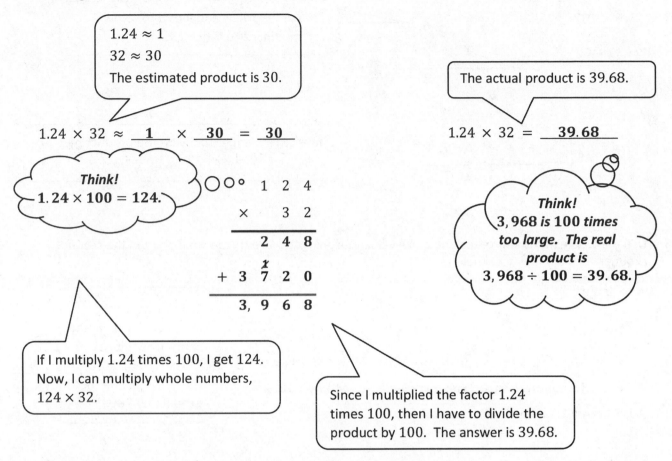

$1.24 \approx 1$
$32 \approx 30$
The estimated product is 30.

$1.24 \times 32 \approx \underline{\ \ 1\ \ } \times \underline{\ \ 30\ \ } = \underline{\ \ 30\ \ }$

The actual product is 39.68.

$1.24 \times 32 = \underline{\ \ 39.68\ \ }$

Think!
$1.24 \times 100 = 124.$

```
        1  2  4
   ×       3  2
   ─────────────
        2  4  8
 +   3  7  2  0
   ─────────────
     3, 9  6  8
```

Think!
$3,968$ is 100 times too large. The real product is
$3,968 \div 100 = 39.68.$

If I multiply 1.24 times 100, I get 124. Now, I can multiply whole numbers, 124×32.

Since I multiplied the factor 1.24 times 100, then I have to divide the product by 100. The answer is 39.68.

2. Solve using the standard algorithm.

    2.46×132

    $= 324.72$

    $$\begin{array}{r} 2\ 4\ 6 \\ \times\ \ 1\ 3\ 2 \\ \hline 4\ 9\ 2 \\ 7\ 3\ 8\ 0 \\ +\ 2\ 4\ 6\ 0\ 0 \\ \hline 3\ 2\ 4\ 7\ 2 \end{array}$$

    > 2.46 times 100 is equal to 246. Now, I can multiply 246 times 132.

    > I have to remember to divide the product by 100.
    > $32,472 \div 100 = 324.72$

3. Use the whole number product and place value reasoning to place the decimal point in the second product. Explain how you know.

    If $54 \times 736 = 39,744$,   then $54 \times 7.36 = \underline{\ \ 397.44\ \ }$.

    > I can compare the factors in both number sentences. Since $736 \div 100 = 7.36$, then I can divide the product by 100.

    7.36 is 736 hundredths, so I can just divide 39,744 by 100.

    $39,744 \div 100 = 397.44$

Lesson 11:   Multiply decimal fractions by multi-digit whole numbers through conversion to a whole number problem and reasoning about the placement of the decimal.
© 2018 Great Minds®. eureka-math.org

EUREKA MATH®

Name _____   Date _____

1. Estimate the product. Solve using the standard algorithm. Use the thought bubbles to show your thinking. (Draw an area model on a separate sheet if it helps you.)

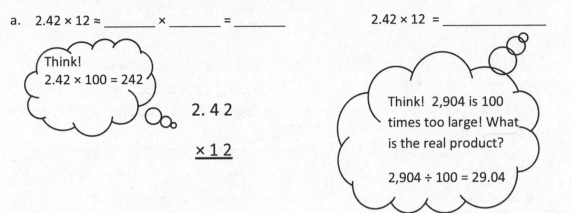

a.   2.42 × 12 ≈ _____ × _____ = _____

Think!
2.42 × 100 = 242

2. 4 2
× 1 2

2.42 × 12 = _____

Think! 2,904 is 100 times too large! What is the real product?

2,904 ÷ 100 = 29.04

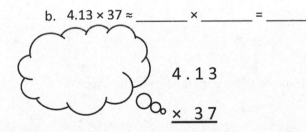

b.   4.13 × 37 ≈ _____ × _____ = _____

4 . 1 3
× 3 7

4.13 × 37 = _____

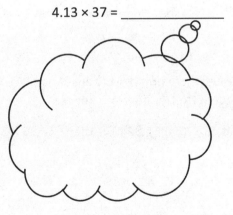

Lesson 11: Multiply decimal fractions by multi-digit whole numbers through conversion to a whole number problem and reasoning about the placement of the decimal.

© 2018 Great Minds®. eureka-math.org

129

2. Solve using the standard algorithm.

   a. 2.03×13

   b. 53.16×34

   c. 371.23×53

   d. 1.57×432

3. Use the whole number product and place value reasoning to place the decimal point in the second product. Explain how you know.

   a. If $36 \times 134 = 4{,}824$ then $36 \times 1.34 =$ _____

   b. If $84 \times 2{,}674 = 224{,}616$ then $84 \times 26.74 =$ _____

   c. $19 \times 3{,}211 = 61{,}009$ then $321.1 \times 19 =$ _____

Lesson 11:   Multiply decimal fractions by multi-digit whole numbers through conversion to a whole number problem and reasoning about the placement of the decimal.
© 2018 Great Minds®. eureka-math.org

4. A slice of pizza costs $1.57. How much will 27 slices cost?

5. A spool of ribbon holds 6.75 meters. A craft club buys 21 spools.

   a. What is the total cost if the ribbon sells for $2 per meter?

   b. If the club uses 76.54 meters to complete a project, how much ribbon will be left?

1. Estimate. Then solve using the standard algorithm. You may draw an area model if it helps you.

$14 \times 3.12 \approx$ __10__ $\times$ __3__ $=$ __30__

$14 \approx 10$
$3.12 \approx 3$
The estimated product is 30.

$$
\begin{array}{r}
3.\,1\ 2 \\
\times\quad 1\ 4 \\
\hline
1\ 2\ 4\ 8 \\
+\ 3\ 1\ 2\ 0 \\
\hline
4\ 3.\,6\ 8
\end{array}
$$

I have to remember to write the product as a number of hundredths.

I'll decompose 14 as $10 + 4$, and 312 hundredths as 300 hundredths + 10 hundredths + 2 hundredths.

	300	+	10	+	2	(hundredths)
4	1,200		40		8	1,248 hundredths
+						
10	3,000		100		20	3,120 hundredths

1,200 hundredths + 40 hundredths + 8 hundredths = 1,248 hundredths.

3,000 hundredths + 100 hundredths + 20 hundredths = 3,120 hundredths.

1,248 hundredths + 3,120 hundredths = 4,368 hundredths, or 43.68.

EUREKA MATH

Lesson 12: Reason about the product of a whole number and a decimal with hundredths using place value understanding and estimation.

133

2 Estimate. Then solve using the standard algorithm.

   a. $0.47 \times 32 \approx$ __**0.5**__ $\times$ __**30**__ $=$ __**15**__

   > $0.47 \approx 0.5$
   > $32 \approx 30$
   > Multiplying 0.5 times 30 is the same as taking half of 30. The estimated product is 15.

   > I'll think of multiplying $0.47 \times 100 = 47$. Now, I'll think of multiplying 47 times 32.

   ```
         0. 4  7
      ×     3  2
      ̶ ̶ ̶ ̶ ̶ ̶ ̶
            9  4
   +   1  4  1  0
      ̶ ̶ ̶ ̶ ̶ ̶ ̶
      1  5. 0  4
   ```

   > I have to remember to write the product as a number of hundredths. $1{,}504 \div 100 = 15.04$.

   b. $6.04 \times 307 \approx$ __**6**__ $\times$ __**300**__ $=$ __**1,800**__

   > $6.04 \approx 6$
   > $307 \approx 300$
   > 6 ones times 3 hundreds is equal to 18 hundreds, or 1,800.

   ```
         6. 0  4
      ×  3  0  7
      ̶ ̶ ̶ ̶ ̶ ̶ ̶
      4  2  2  8
   + 1 8 1  2  0  0
      ̶ ̶ ̶ ̶ ̶ ̶ ̶
   1, 8  5  4. 2  8
   ```

   > The actual product is 1,854.28, which is very close to my estimated product of 1,800.

3. Tatiana walks to the park every afternoon. In the month of August, she walked 2.35 miles each day. How far did Tatiana walk during the month of August?

   There are 31 days in August.

   Tatiana walked 72.85 miles in August.

   > I'll multiply 2.35 times 31 days to find the total distance Tatiana walks during the month of August.

   ```
        2. 3  5
      ×    3  1
      ̶ ̶ ̶ ̶ ̶ ̶
        2  3  5
   +  7  0  5  0
      ̶ ̶ ̶ ̶ ̶ ̶
      7  2. 8  5
   ```

Lesson 12:   Reason about the product of a whole number and a decimal with hundredths using place value understanding and estimation.

Name _____   Date _____

1. Estimate. Then, solve using the standard algorithm. You may draw an area model if it helps you.

    a.   24 × 2.31 ≈ _____ × _____ = _____

$$\begin{array}{r} 2.31 \\ \times\ \ 24 \\ \hline \end{array}$$

    b.   5.42 × 305 ≈ _____ × _____ = _____

$$\begin{array}{r} 5.42 \\ \times 305 \\ \hline \end{array}$$

EUREKA MATH

Lesson 12:   Reason about the product of a whole number and a decimal with hundredths using place value understanding and estimation.

135

© 2018 Great Minds®. eureka-math.org

2. Estimate. Then, solve using the standard algorithm. Use a separate sheet to draw the area model if it helps you.

a. $1.23 \times 21 \approx$ _____ × _____ = _____

b. $3.2 \times 41 \approx$ _____ × _____ = _____

c. $0.32 \times 41 \approx$ _____ × _____ = _____

d. $0.54 \times 62 \approx$ _____ × _____ = _____

e. $6.09 \times 28 \approx$ _____ × _____ = _____

f. $6.83 \times 683 \approx$ _____ × _____ = _____

g. $6.09 \times 208 \approx$ _____ × _____ = _____

h. $171.76 \times 555 \approx$ _____ × _____ = _____

Lesson 12: Reason about the product of a whole number and a decimal with hundredths using place value understanding and estimation.

EUREKA MATH

3. Eric's goal is to walk 2.75 miles to and from the park every day for an entire year. If he meets his goal, how many miles will Eric walk?

4. Art galleries often price paintings by the square inch. If a painting measures 22.5 inches by 34 inches and costs $4.15 per square inch, what is the selling price for the painting?

5. Gerry spends $1.25 each day on lunch at school. On Fridays, she buys an extra snack for $0.55. How much money will she spend in two weeks?

Lesson 12:   Reason about the product of a whole number and a decimal with hundredths using place value understanding and estimation.

137

1. Solve.

    a. Convert years to days.

        5 years = **5 × (1 year)**

                = **5 × (365 days)**

                = **1, 825 days**

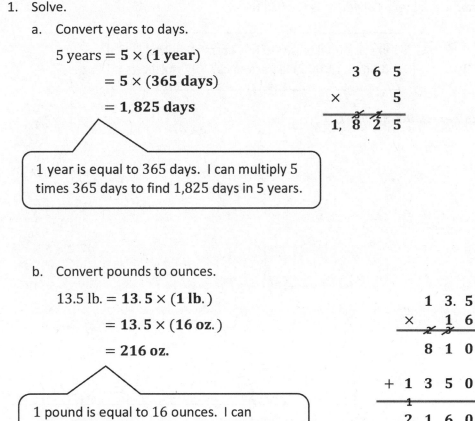

$$\begin{array}{r} 3\ 6\ 5 \\ \times\ \ \ \ \ 5 \\ \hline 1,\ 8\ 2\ 5 \end{array}$$

> 1 year is equal to 365 days. I can multiply 5 times 365 days to find 1,825 days in 5 years.

    b. Convert pounds to ounces.

        13.5 lb. = **13. 5 × (1 lb.)**

                  = **13. 5 × (16 oz.)**

                  = **216 oz.**

$$\begin{array}{r} 1\ 3.\ 5 \\ \times\ \ \ 1\ 6 \\ \hline 8\ 1\ 0 \\ +\ 1\ 3\ 5\ 0 \\ \hline 2\ 1\ 6.\ 0 \end{array}$$

> 1 pound is equal to 16 ounces. I can multiply 13.5 times 16 ounces to find that there are 216 ounces in 13.5 pounds.

2. After solving, write a statement to express each conversion.

    a. The height of a male ostrich is 7.3 meters. What is his height in centimeters?

        7. 3 m = 7. 3 × (1 m)

                  = **7. 3 × (100 cm)**

                  = **730 cm**

> 1 meter is equal to 100 centimeters. I multiply 7.3 times 100 centimeters to get 730 centimeters.

        His height is 730 *centimeters.*

b. The capacity of a container is 0.3 liter. Convert this to milliliters.

$$0.3\text{ L} = 0.3 \times (1\text{ L})$$

$$= 0.3 \times (1,000\text{ ml})$$

$$= 300\text{ ml}$$

> 1 liter is equal to 1,000 milliliters. I multiply 0.3 times 1,000 milliliters to get 300 milliliters.

The capacity of the container is 300 milliliters.

Name _____   Date _____

1. Solve. The first one is done for you.

a. Convert weeks to days. 6 weeks = 6 × (1 week) = 6 × (7 days) = 42 days	b. Convert years to days. 7 years = _____ × (_____ year) = _____ × (_____ days) = _____ days
c. Convert meters to centimeters. 4.5 m = _____ × (_____ m) = _____ × (_____ cm) = _____ cm	d. Convert pounds to ounces. 12.6 pounds
e. Convert kilograms to grams. 3.09 kg	f. Convert yards to inches. 245 yd

Lesson 13:   Use whole number multiplication to express equivalent measurements.

141

© 2018 Great Minds®. eureka-math.org

2. After solving, write a statement to express each conversion. The first one is done for you.

a. Convert the number of hours in a day to minutes. $\quad$ 24 hours $= 24 \times (1\,\text{hour})$ $\qquad\qquad\quad = 24 \times (60\,\text{minutes})$ $\qquad\qquad\quad = 1{,}440\,\text{minutes}$ One day has 24 hours, which is the same as 1,440 minutes.	b. A newborn giraffe weighs about 65 kilograms. How much does it weigh in grams?
c. The average height of a female giraffe is 4.6 meters. What is her height in centimeters?	d. The capacity of a beaker is 0.1 liter. Convert this to milliliters.
e. A pig weighs 9.8 pounds. Convert the pig's weight to ounces.	f. A marker is 0.13 meters long. What is the length in millimeters?

Lesson 13:   Use whole number multiplication to express equivalent measurements.

1. Solve.

   a. Convert quarts to gallons.

   $$28 \text{ quarts} = \mathbf{28 \times (1 \text{ quart})}$$
   $$= \mathbf{28 \times \left(\frac{1}{4} \text{ gallon}\right)}$$
   $$= \mathbf{\frac{28}{4} \text{ gallons}}$$
   $$= \mathbf{7 \text{ gallons}}$$

   > 1 quart is equal to $\frac{1}{4}$ gallon. I multiply 28 times $\frac{1}{4}$ gallon to find 7 gallons is equal to 28 quarts.

   b. Convert grams to kilograms.

   $$5{,}030 \text{ g} = \mathbf{5{,}030 \times (1 \text{ g})}$$
   $$= \mathbf{5{,}030 \times (0.001 \text{ kg})}$$
   $$= \mathbf{5.030 \text{ kg}}$$

   > 1 gram is equal to 0.001 kilogram. I multiply 5,030 times 0.001 kilogram to get 5.030 kilograms.

2. After solving, write a statement to express each conversion.

   a. A jug of milk holds 16 cups. Convert 16 cups to pints.

   $$\mathbf{16 \text{ cups} = 16 \times (1 \text{ cup})}$$
   $$= \mathbf{16 \times \left(\frac{1}{2} \text{ pint}\right)}$$
   $$= \mathbf{\frac{16}{2} \text{ pints}}$$
   $$= \mathbf{8 \text{ pints}}$$

   > 1 cup is equal to $\frac{1}{2}$ pint. I multiply 16 times $\frac{1}{2}$ pint to find that 8 pints is equal to 16 cups.

   16 *cups is equal to 8 pints.*

   b. The length of a table is 305 centimeters. What is its length in meters?

   $$\mathbf{305 \text{ cm} = 305 \times (1 \text{ cm})}$$
   $$= \mathbf{305 \times (0.01 \text{ m})}$$
   $$= \mathbf{3.05 \text{ m}}$$

   > 1 centimeter is equal to 0.01 meter. I multiply 305 times 0.01 meter to get 3.05 meters.

   The table's length is 3.05 meters.

Lesson 14:   Use fraction and decimal multiplication to express equivalent measurements.

143

© 2018 Great Minds®. eureka-math.org

Name _____   Date _____

1.   Solve. The first one is done for you.

a. Convert days to weeks.	b. Convert quarts to gallons.

a.   Convert days to weeks.

$$42 \text{ days} = 42 \times (1 \text{ day})$$

$$= 42 \times \left(\frac{1}{7} \text{ week}\right)$$

$$= \frac{42}{7} \text{ week}$$

$$= 6 \text{ weeks}$$

b.   Convert quarts to gallons.

$$36 \text{ quarts} = \underline{\hspace{1.5cm}} \times (1 \text{ quart})$$

$$= \underline{\hspace{1.5cm}} \times \left(\frac{1}{4} \text{ gallon}\right)$$

$$= \underline{\hspace{1.5cm}} \text{ gallons}$$

$$= \underline{\hspace{1.5cm}} \text{ gallons}$$

c.   Convert centimeters to meters.

$$760 \text{ cm} = \underline{\hspace{1.5cm}} \times (\underline{\hspace{1.5cm}} \text{ cm})$$

$$= \underline{\hspace{1.5cm}} \times (\underline{\hspace{1.5cm}} \text{ m})$$

$$= \underline{\hspace{1.5cm}} \text{ m}$$

d.   Convert meters to kilometers.

$$2{,}485 \text{ m} = \underline{\hspace{1.5cm}} \times (\underline{\hspace{1.5cm}} \text{ m})$$

$$= \underline{\hspace{1.5cm}} \times (0.001 \text{ km})$$

$$= \underline{\hspace{1.5cm}} \text{ km}$$

e.   Convert grams to kilograms.

$$3{,}090 \text{ g} =$$

f.   Convert milliliters to liters.

$$205 \text{ mL} =$$

EUREKA MATH

Lesson 14:   Use fraction and decimal multiplication to express equivalent measurements.

© 2018 Great Minds®. eureka-math.org

145

2. After solving, write a statement to express each conversion. The first one is done for you.

a. The screen measures 36 inches. Convert 36 inches to feet. $\quad$ 36 inches = 36 × (1 inch) $\qquad = 36 \times \left(\dfrac{1}{12} \text{ feet}\right)$ $\qquad = \dfrac{36}{12} \text{ feet}$ $\qquad = 3 \text{ feet}$ The screen measures 36 inches or 3 feet.	b. A jug of juice holds 8 cups. Convert 8 cups to pints.
c. The length of the flower garden is 529 centimeters. What is its length in meters?	d. The capacity of a container is 2,060 milliliters. Convert this to liters.
e. A hippopotamus weighs 1,560,000 grams. Convert the hippopotamus' weight to kilograms.	f. The distance was 372,060 meters. Convert the distance to kilometers.

Lesson 14: $\quad$ Use fraction and decimal multiplication to express equivalent measurements.

1. A bag of peanuts is 5 times as heavy as a bag of sunflower seeds. The bag of peanuts also weighs 920 grams more than the bag of sunflower seeds.

    a. What is the total weight in grams for the bag of peanuts and the bag of sunflower seeds?

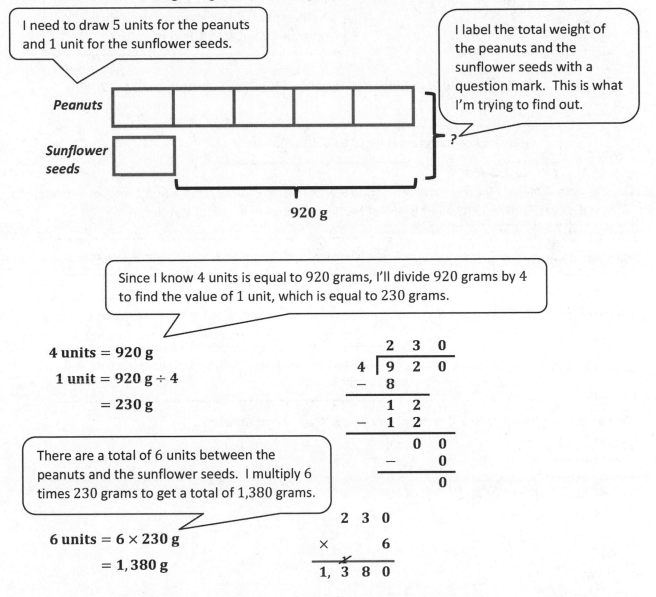

I need to draw 5 units for the peanuts and 1 unit for the sunflower seeds.

I label the total weight of the peanuts and the sunflower seeds with a question mark. This is what I'm trying to find out.

Peanuts

Sunflower seeds

?

920 g

Since I know 4 units is equal to 920 grams, I'll divide 920 grams by 4 to find the value of 1 unit, which is equal to 230 grams.

4 units = 920 g

1 unit = 920 g ÷ 4

= 230 g

There are a total of 6 units between the peanuts and the sunflower seeds. I multiply 6 times 230 grams to get a total of 1,380 grams.

6 units = 6 × 230 g

= 1,380 g

$$\begin{array}{r} 2\ 3\ 0 \\ 4\overline{)9\ 2\ 0} \\ -\ 8 \\ \hline 1\ 2 \\ -\ 1\ 2 \\ \hline 0\ 0 \\ -\ 0 \\ \hline 0 \end{array}$$

$$\begin{array}{r} 2\ 3\ 0 \\ \times\ 6 \\ \hline 1,3\ 8\ 0 \end{array}$$

The total weight for the bag of peanuts and the bag of sunflower seeds is 1,380 grams.

Lesson 15: Solve two-step word problems involving measurement conversions.

147

EUREKA MATH

b. Express the total weight of the bag of peanuts and the bag of sunflower seeds in kilograms.

$1,380 \text{ g} = 1,380 \times (1 \text{ g})$

$\qquad = 1,380 \times (0.001 \text{ kg})$

$\qquad = 1.380 \text{ kg}$

> 1 gram is equal to 0.001 kilogram. I multiply 1,380 times 0.001 kilogram to find that 1.38 kilograms is equal to 1,380 grams.

The total weight of the bag of peanuts and the bag of sunflower seeds is **1.38** **kilograms.**

> 4 meters 50 centimeters is equal to 450 centimeters.

2. Gabriel cut a 4 meter 50 centimeter string into 9 equal pieces. Michael cut a 508 centimeter string into 10 equal pieces. How much longer is one of Michael's strings than one of Gabriel's?

Gabriel: $450 \text{ cm} \div 9 = 50 \text{ cm}$

> Each piece of Gabriel's string is 50 centimeters long.

Michael: $508 \text{ cm} \div 10 = 50.8 \text{ cm}$

> Each piece of Michael's string is 50.8 centimeters long.

$50.8 \text{ cm} - 50 \text{ cm} = 0.8 \text{ cm}$

> I'll subtract to find the difference between Michael and Gabriel's strings.

One of Michael's strings is **0.8** **centimeters longer than one of Gabriel's.**

Lesson 15:   Solve two-step word problems involving measurement conversions.

Name _____   Date _____

Solve.

1. Tia cut a 4-meter 8-centimeter wire into 10 equal pieces. Marta cut a 540-centimeter wire into 9 equal pieces. How much longer is one of Marta's wires than one of Tia's?

2. Jay needs 19 quarts more paint for the outside of his barn than for the inside. If he uses 107 quarts in all, how many gallons of paint will be used to paint the inside of the barn?

3. String A is 35 centimeters long. String B is 5 times as long as String A. Both are necessary to create a decorative bottle. Find the total length of string needed for 17 identical decorative bottles. Express your answer in meters.

4. A pineapple is 7 times as heavy as an orange. The pineapple also weighs 870 grams more than the orange.

   a. What is the total weight in grams for the pineapple and orange?

   b. Express the total weight of the pineapple and orange in kilograms.

1. Divide. Draw place value disks to show your thinking for (a).

   a. $400 \div 10 = \mathbf{40}$

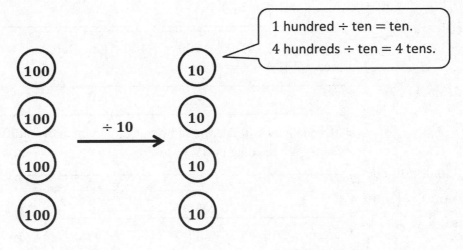

> 1 hundred ÷ ten = ten.
> 4 hundreds ÷ ten = 4 tens.

   b. $650,000 \div 100$

   $= \mathbf{6,500 \div 1}$

   $= \mathbf{6,500}$

   > I can divide both the dividend and the divisor by 100, so I can rewrite the division sentence as 6,500 ÷ 1. The answer is 6,500.

> Dividing by 40 is the same thing as dividing by 10 and then dividing by 4.

2. Divide.

   a. $240,000 \div 40$

   $= \mathbf{240,000 \div 10 \div 4}$

   $= \mathbf{24,000 \div 4}$

   $= \mathbf{6,000}$

   > I can solve 240,000 ÷ 10 = 24,000. Then I can find that 24,000 ÷ 4 = 6,000.

   > In unit form, this is 24 thousands ÷ 4 = 6 thousands.

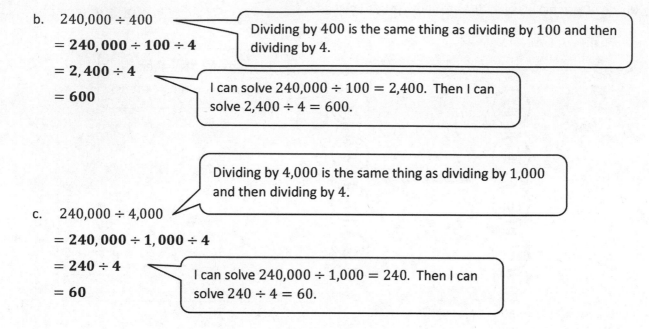

b. $240,000 \div 400$

$= 240,000 \div 100 \div 4$

$= 2,400 \div 4$

$= 600$

Dividing by 400 is the same thing as dividing by 100 and then dividing by 4.

I can solve $240,000 \div 100 = 2,400$. Then I can solve $2,400 \div 4 = 600$.

Dividing by 4,000 is the same thing as dividing by 1,000 and then dividing by 4.

c. $240,000 \div 4,000$

$= 240,000 \div 1,000 \div 4$

$= 240 \div 4$

$= 60$

I can solve $240,000 \div 1,000 = 240$. Then I can solve $240 \div 4 = 60$.

Lesson 16: Use *divide by 10* patterns for multi-digit whole number division.

EUREKA MATH®

Name _____ Date _____

1. Divide. ~~Draw place value disks to show your thinking for (a) and (c). You may draw disks on your personal white board to solve the others if necessary.~~

a. 300 ÷ 10 30	b. 450 ÷ 10 45
c. 18,000 ÷ 100 180	d. 730,000 ÷ 100 7,300
e. 900,000 ÷ 1,000 900	f. 680,000 ÷ 1,000 680

2. Divide. The first one is done for you.

a. 18,000 ÷ 20 = 18,000 ÷ 10 ÷ 2 = 1,800 ÷ 2 = 900	b. 18,000 ÷ 200	c. 18,000 ÷ 2,000
d. 420,000 ÷ 60	e. 420,000 ÷ 600	f. 420,000 ÷ 6,000
g. 24,000 ÷ 30	h. 560,000 ÷ 700	i. 450,000 ÷ 9,000

Lesson 16:   *Use divide by 10 patterns for multi-digit whole number division.*

3. A stadium holds 50,000 people. The stadium is divided into 250 different seating sections. How many seats are in each section?

4. Over the course of a year, a tractor trailer commutes 160,000 miles across America.

a. Assuming a trucker changes his tires every 40,000 miles, and that he starts with a brand new set of tires, how many sets of tires will he use in a year?

b. If the trucker changes the oil every 10,000 miles, and he starts the year with a fresh oil change, how many times will he change the oil in a year?

1. Estimate the quotient for the following problems.

> I look at the divisor, 33, and round it to the nearest ten. $33 \approx 30$

a. $612 \div 33$

> I need to think of a multiple of 30 that's closest to 612. 600 works.

$\approx 600 \div 30$

$= 20$

> I use the simple fact, $6 \div 3 = 2$, to help me solve $600 \div 30 = 20$.

> I look at the divisor, 78, and round it to the nearest ten. $78 \approx 80$

b. $735 \div 78$

> I'll think of a multiple of 80 that is close to 735. 720 is the closest multiple.

$\approx 720 \div 80$

$= 9$

> I use the simple fact, $72 \div 8 = 9$, to help me solve $720 \div 80 = 9$.

> I look at the divisor, 99, and round to the nearest ten. $99 \approx 100$

c. $821 \div 99$

> I can think of a multiple of 100 that is close to 821. 800 is the closest multiple.

$\approx 800 \div 100$

$= 8$

> I can use the simple fact, $8 \div 1 = 8$, to help solve $800 \div 100 = 8$.

EUREKA MATH®
© 2018 Great Minds®. eureka-math.org

2. A baker spent $989 buying 48 pounds of nuts. About how much does each pound of nuts cost?

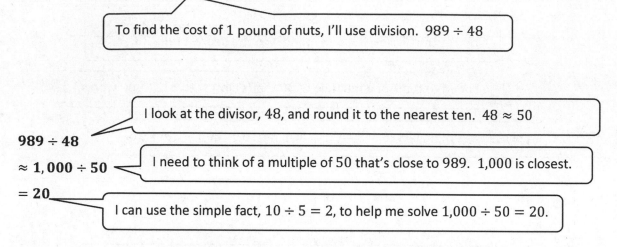

To find the cost of 1 pound of nuts, I'll use division. 989 ÷ 48

I look at the divisor, 48, and round it to the nearest ten. 48 ≈ 50

989 ÷ 48

≈ 1,000 ÷ 50

I need to think of a multiple of 50 that's close to 989. 1,000 is closest.

= 20

I can use the simple fact, 10 ÷ 5 = 2, to help me solve 1,000 ÷ 50 = 20.

Each pound of nuts costs about $20.

Lesson 17:   Use basic facts to approximate quotients with two-digit divisors.

Name _____ Date _____

1. Estimate the quotient for the following problems. The first one is done for you.

a. 821 ÷ 41 ≈ 800 ÷ 40 = 20	b. 617 ÷ 23 ≈ _____ ÷ _____ = _____	c. 821 ÷ 39 ≈ _____ ÷ _____ = _____
d. 482 ÷ 52 ≈ _____ ÷ _____ = _____	e. 531 ÷ 48 ≈ _____ ÷ _____ = _____	f. 141 ÷ 73 ≈ _____ ÷ _____ = _____
g. 476 ÷ 81 ≈ _____ ÷ _____ = _____	h. 645 ÷ 69 ≈ _____ ÷ _____ = _____	i. 599 ÷ 99 ≈ _____ ÷ _____ = _____
j. 301 ÷ 26 ≈ _____ ÷ _____ = _____	k. 729 ÷ 81 ≈ _____ ÷ _____ = _____	l. 636 ÷ 25 ≈ _____ ÷ _____ = _____
m. 835 ÷ 89 ≈ _____ ÷ _____ = _____	n. 345 ÷ 72 ≈ _____ ÷ _____ = _____	o. 559 ÷ 11 ≈ _____ ÷ _____ = _____

Lesson 17:   Use basic facts to approximate quotients with two-digit divisors.

159

2. Mrs. Johnson spent $611 buying lunch for 78 students. If all the lunches cost the same, about how much did she spend on each lunch?

3. An oil well produces 172 gallons of oil every day. A standard oil barrel holds 42 gallons of oil. About how many barrels of oil will the well produce in one day? Explain your thinking.

Lesson 17:   Use basic facts to approximate quotients with two-digit divisors.

1. Estimate the quotients for the following problems.

a. $3,782 \div 23$

$\approx 4,000 \div 20$

$= 200$

> I look at the divisor, 23, and round it to the nearest ten. $23 \approx 20$

> I need to think of a multiple of 20 that's closest to 3,782. 4,000 is closest.

> I use the simple fact, $4 \div 2 = 2$, and unit form to help me solve.
> 4 thousands $\div$ 2 tens = 2 hundreds

b. $2,519 \div 43$

$\approx 2,400 \div 40$

$= 60$

> I look at the divisor, 43, and round to the nearest ten. $43 \approx 40$

> I need to think of a multiple of 40 that's close to 2,519. 2,400 is closest.

> I can use the simple fact, $24 \div 4 = 6$, to help me solve $2,400 \div 40 = 60$.

c. $4,621 \div 94$

$\approx 4,500 \div 90$

$= 50$

> I look at the divisor, 94, and round it to the nearest ten. $94 \approx 90$

> 4,500 is close to 4,621 and is a multiple of 90.

> I can use the simple fact, $45 \div 9 = 5$, to help me solve $4,500 \div 90 = 50$.

Lesson 18: Use basic facts to approximate quotients with two-digit divisors.

161

2. Meilin has saved \$4,825. If she is paid \$68 an hour, about how many hours did she work?

I'll use division to find the number of hours that Meilin worked to save \$4,825.

The divisor, 68, rounds to 70. $68 \approx 70$

$4,825 \div 68$

$\approx 4,900 \div 70$

I need to find a multiple of 70 that's closest to 4,825. 4,900 is closest.

$= 70$

I can use the basic fact, $49 \div 7 = 7$, to help me solve $4,900 \div 70 = 70$.

Meilin worked about 70 hours.

Lesson 18: Use basic facts to approximate quotients with two-digit divisors.

EUREKA MATH

Name _____ Date _____

1. Estimate the quotients for the following problems. The first one is done for you.

a. 8,328 ÷ 41 ≈ 8,000 ÷ 40 = 200	b. 2,109 ÷ 23 ≈ _____ ÷ _____ = _____	c. 8,215 ÷ 38 ≈ _____ ÷ _____ = _____
d. 3,861 ÷ 59 ≈ _____ ÷ _____ = _____	e. 2,899 ÷ 66 ≈ _____ ÷ _____ = _____	f. 5,576 ÷ 92 ≈ _____ ÷ _____ = _____
g. 5,086 ÷ 73 ≈ _____ ÷ _____ = _____	h. 8,432 ÷ 81 ≈ _____ ÷ _____ = _____	i. 9,032 ÷ 89 ≈ _____ ÷ _____ = _____
j. 2,759 ÷ 48 ≈ _____ ÷ _____ = _____	k. 8,194 ÷ 91 ≈ _____ ÷ _____ = _____	l. 4,368 ÷ 63 ≈ _____ ÷ _____ = _____
m. 6,537 ÷ 74 ≈ _____ ÷ _____ = _____	n. 4,998 ÷ 48 ≈ _____ ÷ _____ = _____	o. 6,106 ÷ 25 ≈ _____ ÷ _____ = _____

© 2018 Great Minds®. eureka-math.org

2. 91 boxes of apples hold a total of 2,605 apples. Assuming each box has about the same number of apples, estimate the number of apples in each box.

3. A wild tiger can eat up to 55 pounds of meat in a day. About how many days would it take for a tiger to eat the following prey?

Prey	Weight of Prey	Number of Days
Eland Antelope	1,754 pounds	
Boar	661 pounds	
Chital Deer	183 pounds	
Water Buffalo	2,322 pounds	

1. Divide, and then check.

a. $87 \div 40$

I use the estimation strategy from the previous lesson to help me solve. $80 \div 40 = 2$. The estimated quotient is 2.

I write the remainder of 7 here next to the quotient of 2.

I check my answer by multiplying the divisor of 40 by the quotient of 2 and then add the remainder of 7.

$$\begin{array}{r} 2 \ \ R\,7 \\ 40\overline{\smash{\big)}\,8\ 7} \\ -\ \ 8\ 0 \\ \hline 7 \end{array}$$

2 groups of 40 is equal to 80.

Check:

$40 \times 2 = 80$

$80 + 7 = 87$

The difference between 87 and 80 is 7.

This 87 matches the original dividend in the problem, which means I divided correctly. The quotient is 2 with a remainder of 7.

b. $451 \div 70$

I estimate to find the quotient. $420 \div 70 = 6$

The quotient is 6 with a remainder of 31.

After checking, I see that 451 does match the original dividend in the problem.

$$\begin{array}{r} 6 \ \ R\,31 \\ 70\overline{\smash{\big)}\,4\ 5\ 1} \\ -\ \ 4\ 2\ 0 \\ \hline 3\ 1 \end{array}$$

Check:

$70 \times 6 = 420$

$420 + 31 = 451$

The quotient is 6 with a remainder of 31.

EUREKA MATH®

Lesson 19:   Divide two- and three-digit dividends by multiples of 10 with single-digit quotients, and make connections to a written method.

165

© 2018 Great Minds®. eureka-math.org

2. How many groups of thirty are in two hundred twenty-four?

I use division to find how many 30's are in 224. But first, I estimate to find the quotient. $210 \div 30 = 7$

There are 7 groups of thirty in 224 with a remainder of 14.

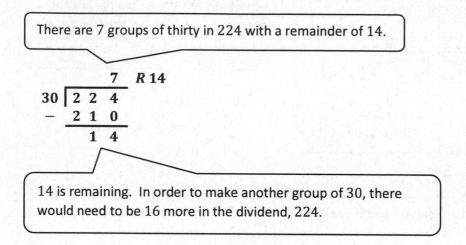

$$
\begin{array}{r}
\phantom{30\,\overline{\,}}7\ \ R\,14 \\
30\,\overline{\big)\,2\ \ 2\ \ 4} \\
-\ \ 2\ \ 1\ \ 0 \\
\hline
1\ \ 4
\end{array}
$$

14 is remaining. In order to make another group of 30, there would need to be 16 more in the dividend, 224.

There are 7 groups of thirty in two hundred twenty-four.

Lesson 19: Divide two- and three-digit dividends by multiples of 10 with single-digit quotients, and make connections to a written method.

EUREKA MATH®

Name _____ Date _____

1. Divide, and then check using multiplication. The first one is done for you.

   a. $71 \div 20$

   $$
   \begin{array}{r}
   3 \quad \text{R } 11 \\
   2\,0\,\overline{\smash{)}7\quad 1} \\
   -\ 6\quad 0 \\
   \hline
   1\quad 1
   \end{array}
   $$

   Check:

   $20 \times 3 = 60$

   $60 + 11 = 71$

   b. $90 \div 40$

   c. $95 \div 60$

   d. $280 \div 30$

   e. $437 \div 60$

   f. $346 \div 80$

Lesson 19: Divide two- and three-digit dividends by multiples of 10 with single-digit quotients, and make connections to a written method.

© 2018 Great Minds®. eureka-math.org

167

2. A number divided by 40 has a quotient of 6 with a remainder of 16. Find the number.

3. A shipment of 288 reams of paper was delivered. Each of the 30 classrooms received an equal share of the paper. Any extra reams of paper were stored. After the paper was distributed to the classrooms, how many reams of paper were stored?

4. How many groups of sixty are in two hundred forty-four?

Lesson 19:   Divide two- and three-digit dividends by multiples of 10 with single-digit quotients, and make connections to a written method.

EUREKA MATH®

1. Divide. Then check with multiplication

a. $48 \div 21$

> I do a quick mental estimation to find the quotient.
> $40 \div 20 = 2$

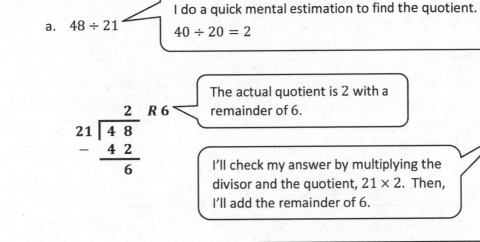

> The actual quotient is 2 with a remainder of 6.

$$\begin{array}{r} 2 R\,6 \\ 21\overline{)\,4\ 8} \\ -\ 4\ 2 \\ \hline 6 \end{array}$$

> I'll check my answer by multiplying the divisor and the quotient, 21×2. Then, I'll add the remainder of 6.

Check:

$$\begin{array}{r} 2\ 1 \\ \times\ \ 2 \\ \hline 4\ 2 \end{array} \qquad \begin{array}{r} 4\ 2 \\ +\ \ \ 6 \\ \hline 4\ 8 \end{array}$$

> This 48 matches the original dividend in the problem, which means I divided correctly. The quotient is 2 with a remainder of 6.

b. $79 \div 38$

> I do a quick mental estimation to find the quotient.
> $80 \div 40 = 2$

$$\begin{array}{r} 2 R\,3 \\ 38\overline{)\,7\ 9} \\ -\ 7\ 6 \\ \hline 3 \end{array}$$

> The actual quotient is 2 with a remainder of 3.

Check:

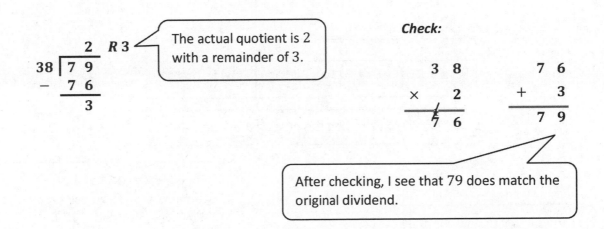

$$\begin{array}{r} 3\ 8 \\ \times\ \ 2 \\ \hline 7\ 6 \end{array} \qquad \begin{array}{r} 7\ 6 \\ +\ \ \ 3 \\ \hline 7\ 9 \end{array}$$

> After checking, I see that 79 does match the original dividend.

Lesson 20:   Divide two- and three-digit dividends by two-digit divisors with single-digit quotients, and make connections to a written method.

169

© 2018 Great Minds®. eureka-math.org

Area is equal to length times width. So, I can use the area divided by the length to find the width.

$A = l \times w$ and $A \div l = w$

2. A rectangular 95-square-foot vegetable garden has a length of 19 feet. What is the width of the vegetable garden?

$95 \div 19 = 5$

I'll do a quick mental estimation to help me solve.

$100 \div 20 = 5$

The quotient of 5 means the width is 5 feet, with 0 feet remaining.

$$\begin{array}{r} 5 \\ 19\overline{)9\ 5} \\ -\ 9\ 5 \\ \hline 0 \end{array}$$

The width of the vegetable garden is 5 feet.

3. A number divided by 41 has a quotient of 4 with 15 as a remainder. Find the number.

In other words, 4 units of 41, plus 15 more, is equal to what number?

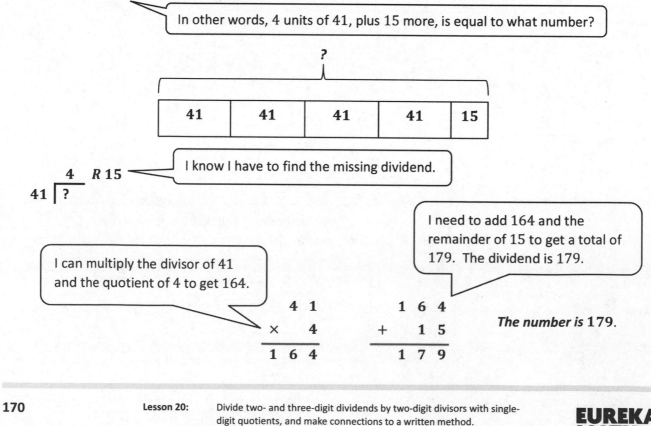

?

| 41 | 41 | 41 | 41 | 15 |

I know I have to find the missing dividend.

$$\begin{array}{r} 4 \quad R\ 15 \\ 41\overline{)?} \end{array}$$

I need to add 164 and the remainder of 15 to get a total of 179. The dividend is 179.

I can multiply the divisor of 41 and the quotient of 4 to get 164.

$$\begin{array}{r} 4\ 1 \\ \times\ \ \ 4 \\ \hline 1\ 6\ 4 \end{array} \qquad \begin{array}{r} 1\ 6\ 4 \\ +\ \ 1\ 5 \\ \hline 1\ 7\ 9 \end{array}$$

The number is 179.

Lesson 20:   Divide two- and three-digit dividends by two-digit divisors with single-digit quotients, and make connections to a written method.

© 2018 Great Minds®. eureka-math.org

EUREKA
MATH®

Name _____ Date _____

1. Divide. Then, check with multiplication. The first one is done for you.

   a. 72 ÷ 31

   b. 89 ÷ 21

```
         2 R 10
   3 1 | 7 2
       - 6 2
         1 0
```

   Check:

   $31 \times 2 = 62$

   $62 + 10 = 72$

   c. 94 ÷ 33

   d. 67 ÷ 19

   e. 79 ÷ 25

   f. 83 ÷ 21

Lesson 20:   Divide two- and three-digit dividends by two-digit divisors with single-digit quotients, and make connections to a written method.

171

2. A 91 square foot bathroom has a length of 13 feet. What is the width of the bathroom?

3. While preparing for a morning conference, Principal Corsetti is laying out 8 dozen bagels on square plates.
    Each plate can hold 14 bagels.

    a. How many plates of bagels will Mr. Corsetti have?

    b. How many more bagels would be needed to fill the final plate with bagels?

Lesson 20:   Divide two- and three-digit dividends by two-digit divisors with single-
                        digit quotients, and make connections to a written method.

        © 2018 Great Minds®. eureka-math.org

1. Divide. Then check using multiplication.

a. $235 \div 68$

> I can find the estimated quotient and then divide using the long division algorithm.

> I can estimate to find the quotient. $210 \div 70 = 3$

> I'll use the quotient of 3. 3 groups of 68 is 204, and the difference between 235 and 204 is 31. The remainder is 31.

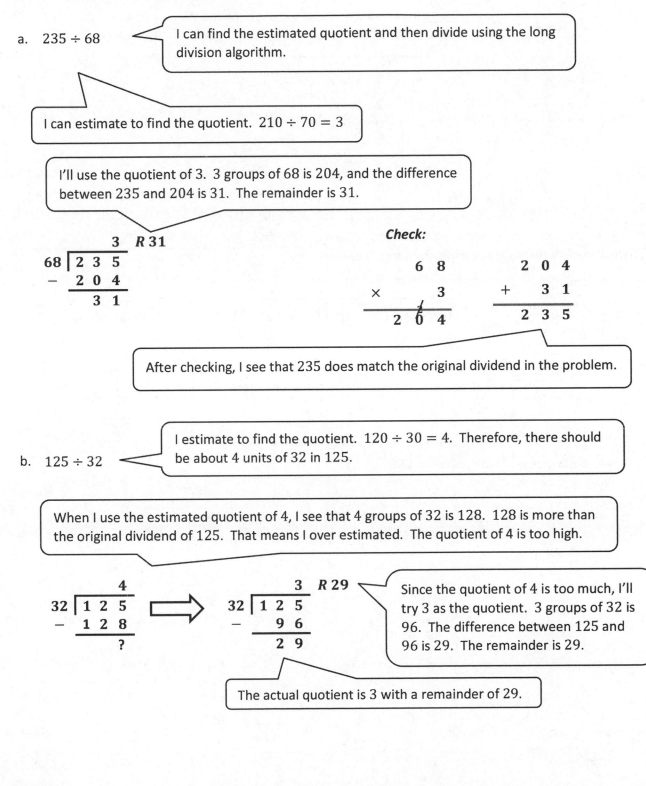

Check:

```
        3  R 31
 68 | 2 3 5
   -  2 0 4
        3 1
```

```
      6 8
   ×     3
   ─────────
      2 0 4
```

```
      2 0 4
   +   3 1
   ─────────
      2 3 5
```

> After checking, I see that 235 does match the original dividend in the problem.

b. $125 \div 32$

> I estimate to find the quotient. $120 \div 30 = 4$. Therefore, there should be about 4 units of 32 in 125.

> When I use the estimated quotient of 4, I see that 4 groups of 32 is 128. 128 is more than the original dividend of 125. That means I over estimated. The quotient of 4 is too high.

```
          4
 32 | 1 2 5
    - 1 2 8
        ?
```
⟹
```
          3  R 29
 32 | 1 2 5
    -   9 6
        2 9
```

> Since the quotient of 4 is too much, I'll try 3 as the quotient. 3 groups of 32 is 96. The difference between 125 and 96 is 29. The remainder is 29.

> The actual quotient is 3 with a remainder of 29.

 EUREKA MATH

Lesson 21: Divide two- and three-digit dividends by two-digit divisors with single-digit quotients, and make connections to a written method.

173

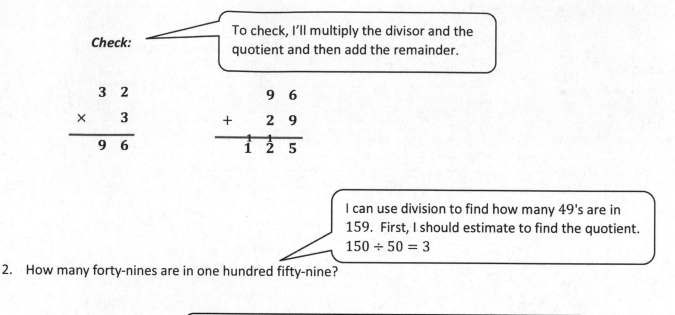

Check: To check, I'll multiply the divisor and the quotient and then add the remainder.

$$
\begin{array}{r}
3\ 2 \\
\times\quad 3 \\
\hline
9\ 6
\end{array}
\qquad
\begin{array}{r}
9\ 6 \\
+\quad 2\ 9 \\
\hline
1\ 2\ 5
\end{array}
$$

I can use division to find how many 49's are in 159. First, I should estimate to find the quotient.
$150 \div 50 = 3$

2. How many forty-nines are in one hundred fifty-nine?

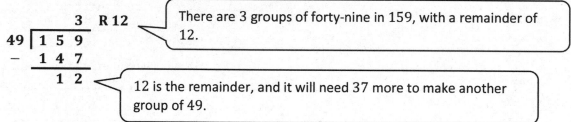

$$
\begin{array}{r}
3\ \ \text{R}\,12 \\
49\,\overline{)\,1\ 5\ 9} \\
-\ 1\ 4\ 7 \\
\hline
1\ 2
\end{array}
$$

There are 3 groups of forty-nine in 159, with a remainder of 12.

12 is the remainder, and it will need 37 more to make another group of 49.

There are 3 groups of forty-nine in 159.

Lesson 21: Divide two- and three-digit dividends by two-digit divisors with single-digit quotients, and make connections to a written method.

EUREKA MATH

Name _____   Date _____

1. Divide. Then, check using multiplication. The first one is done for you.

   a.   129 ÷ 21

   $$\begin{array}{r} 6 \text{ R } 3 \\ 21 \overline{\smash{\big)}\ 1\ 2\ 9} \\ -\ \underline{1\ 2\ 6} \\ 3 \end{array}$$

   Check:

   21 × 6 = 126

   126 + 3 = 129

   b.   158 ÷ 37

   c.   261 ÷ 49

   d.   574 ÷ 82

Lesson 21:   Divide two- and three-digit dividends by two-digit divisors with single-digit quotients, and make connections to a written method.

© 2018 Great Minds®. eureka-math.org

175

e.   464 ÷ 58

f.   640 ÷ 79

2.   It takes Juwan exactly 35 minutes by car to get to his grandmother's. The nearest parking area is a 4-minute walk from her apartment. One week, he realized that he spent 5 hours and 12 minutes traveling to her apartment and then back home. How many round trips did he make to visit his grandmother?

3.   How many eighty-fours are in 672?

Lesson 21:   Divide two- and three-digit dividends by two-digit divisors with single-digit quotients, and make connections to a written method.

1. Divide. Then check using multiplication.

   a. $874 \div 41$

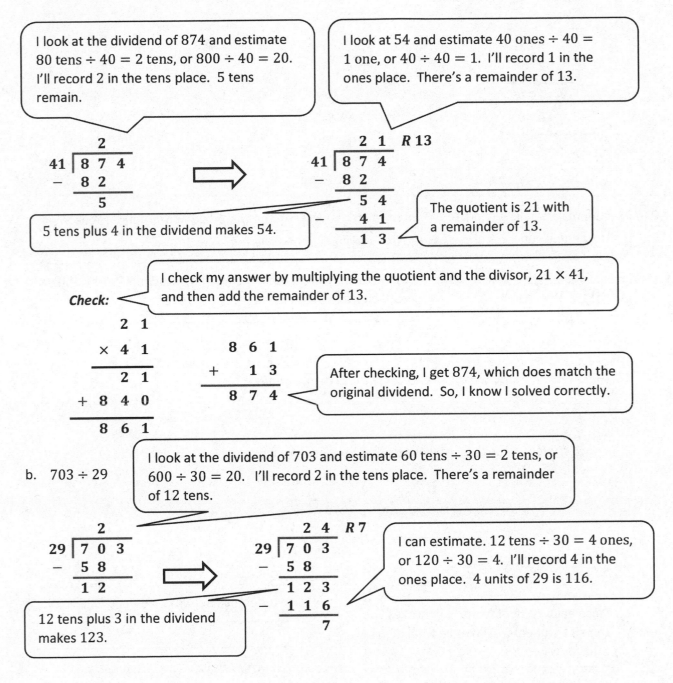

I look at the dividend of 874 and estimate 80 tens ÷ 40 = 2 tens, or 800 ÷ 40 = 20. I'll record 2 in the tens place. 5 tens remain.

I look at 54 and estimate 40 ones ÷ 40 = 1 one, or 40 ÷ 40 = 1. I'll record 1 in the ones place. There's a remainder of 13.

5 tens plus 4 in the dividend makes 54.

The quotient is 21 with a remainder of 13.

I check my answer by multiplying the quotient and the divisor, 21 × 41, and then add the remainder of 13.

After checking, I get 874, which does match the original dividend. So, I know I solved correctly.

   b. $703 \div 29$

I look at the dividend of 703 and estimate 60 tens ÷ 30 = 2 tens, or 600 ÷ 30 = 20. I'll record 2 in the tens place. There's a remainder of 12 tens.

I can estimate. 12 tens ÷ 30 = 4 ones, or 120 ÷ 30 = 4. I'll record 4 in the ones place. 4 units of 29 is 116.

12 tens plus 3 in the dividend makes 123.

Lesson 22: Divide three- and four-digit dividends by two-digit divisors resulting in two- and three-digit quotients, reasoning about the decomposition of successive remainders in each place value.

© 2018 Great Minds®. eureka-math.org

177

Check: I check my answer by multiplying the quotient and the divisor, and then I add the remainder.

$$
\begin{array}{r}
2\ 4 \\
\times\ 2\ 9 \\
\hline
2\ 1\ 6 \\
+\ 4\ 8\ 0 \\
\hline
6\ 9\ 6
\end{array}
$$

$$
\begin{array}{r}
6\ 9\ 6 \\
+\ \ 0\ 7 \\
\hline
7\ 0\ 3
\end{array}
$$

2. 31 students are selling cupcakes. There are 167 cupcakes to be shared equally among students.

   a. How many cupcakes are left over after sharing them equally?

$$
\begin{array}{r}
5\ \text{R}\,12 \\
31\,\overline{)1\ 6\ 7} \\
-\ 1\ 5\ 5 \\
\hline
1\ 2
\end{array}
$$

167 cupcakes shared equally among 31 students: each student gets 5 cupcakes, with 12 cupcakes left over.

   There are 12 cupcakes left over after sharing them equally.

   b. If each student needs 6 cupcakes to sell, how many more cupcakes are needed?

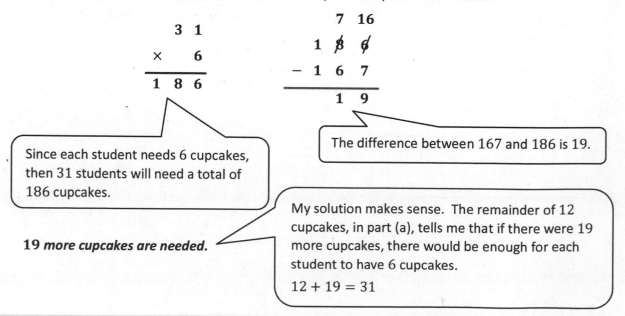

$$
\begin{array}{r}
3\ 1 \\
\times\ \ \ 6 \\
\hline
1\ 8\ 6
\end{array}
$$

$$
\begin{array}{r}
7\ \ 16 \\
1\ \cancel{8}\ \cancel{6} \\
-\ 1\ 6\ 7 \\
\hline
1\ 9
\end{array}
$$

The difference between 167 and 186 is 19.

Since each student needs 6 cupcakes, then 31 students will need a total of 186 cupcakes.

My solution makes sense. The remainder of 12 cupcakes, in part (a), tells me that if there were 19 more cupcakes, there would be enough for each student to have 6 cupcakes.

$12 + 19 = 31$

19 more cupcakes are needed.

Lesson 22:   Divide three- and four-digit dividends by two-digit divisors resulting in two- and three-digit quotients, reasoning about the decomposition of successive remainders in each place value.

© 2018 Great Minds®. eureka-math.org

EUREKA MATH®

Name _Adrienne_ ✓   Date _____

1. Divide. Then, check using multiplication. The first one is done for you.

   a. 487 ÷ 21

   $$
   \begin{array}{r}
   2\ 3\ R\ 4 \\
   21\overline{)4\ 8\ 7} \\
   -\ 4\ 2 \\
   \hline
   6\ 7 \\
   -\ 6\ 3 \\
   \hline
   4
   \end{array}
   $$

   Check:

   21 × 23 = 483

   483 + 4 = 487

   b. 485 ÷ 15

   32 R5

   15)485
   45
   ___
   035
   -30
   ___
   005

   c. 700 ÷ 21

   33 R13

   21)700
   -63
   ___
   070
   -63
   ___
   8

   700

   6 10
   7 0
   -63
   ___
   08

   21
   × 3
   ___
   63

   15
   × 2
   ___
   30

   d. 399 ÷ 31

EUREKA MATH

Lesson 22: Divide three- and four-digit dividends by two-digit divisors resulting in two- and three-digit quotients, reasoning about the decomposition of successive remainders in each place value.

© 2018 Great Minds®. eureka-math.org

179

e. 820 ÷ 42

20

$$
\begin{array}{r}
19 \\
42\overline{)820} \\
-42 \\
\end{array}
$$

$19 \times 42 = 7981$

$798 + 22 = 820$

f. 908 ÷ 56

$1{,}000 \div 50 \approx 20$

$$
56\overline{)908} \quad 16\ R12 \quad \text{check}
$$

$16 \times 56 = 896$

$896 + 12 = 908$

$$
\begin{array}{r}
16\ R12 \\
56\overline{)908} \\
56 \\
\hline
348 \\
336 \\
\hline
012 \\
\end{array}
$$

2. When dividing 878 by 31, a student finds a quotient of 28 with a remainder of 11. Check the student's work, and use the check to find the error in the solution.

3. A baker was going to arrange 432 desserts into rows of 28. The baker divides 432 by 28 and gets a quotient of 15 with remainder 12. Explain what the quotient and remainder represent.

Lesson 22:   Divide three- and four-digit dividends by two-digit divisors resulting in two- and three-digit quotients, reasoning about the decomposition of successive remainders in each place value.

© 2018 Great Minds®. eureka-math.org

1. Divide. Then check using multiplication.

   a. 4,753 ÷ 22

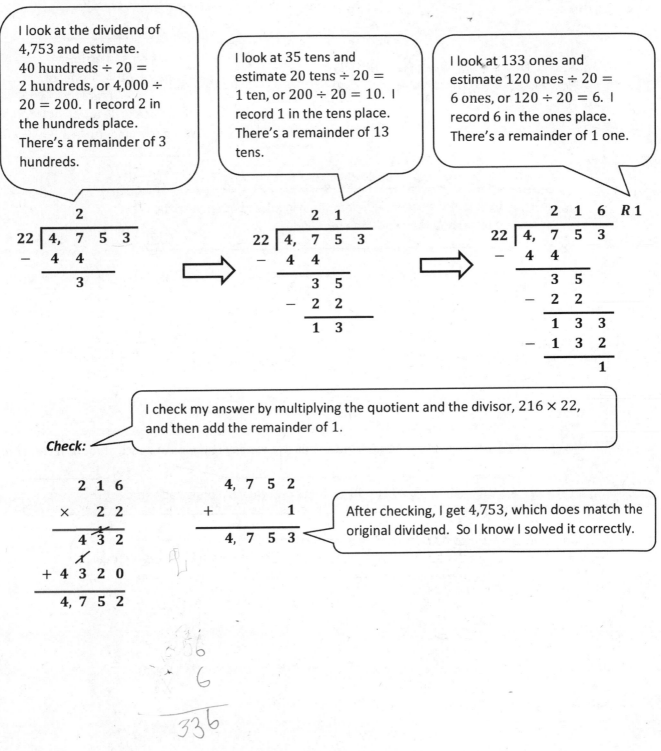

I look at the dividend of 4,753 and estimate. 40 hundreds ÷ 20 = 2 hundreds, or 4,000 ÷ 20 = 200. I record 2 in the hundreds place. There's a remainder of 3 hundreds.

I look at 35 tens and estimate 20 tens ÷ 20 = 1 ten, or 200 ÷ 20 = 10. I record 1 in the tens place. There's a remainder of 13 tens.

I look at 133 ones and estimate 120 ones ÷ 20 = 6 ones, or 120 ÷ 20 = 6. I record 6 in the ones place. There's a remainder of 1 one.

I check my answer by multiplying the quotient and the divisor, 216 × 22, and then add the remainder of 1.

Check:

After checking, I get 4,753, which does match the original dividend. So I know I solved it correctly.

EUREKA MATH®

Lesson 23: Divide three- and four-digit dividends by two-digit divisors resulting in two- and three-digit quotients, reasoning about the decomposition of successive remainders in each place value.

© 2018 Great Minds®. eureka-math.org

181

I look at the dividend of 3,795 and estimate 360 tens ÷ 60 = 6 tens, or 3600 ÷ 60 = 60. I record 6 in the tens place. There's a remainder of 7 tens.

b.   3,795 ÷ 62

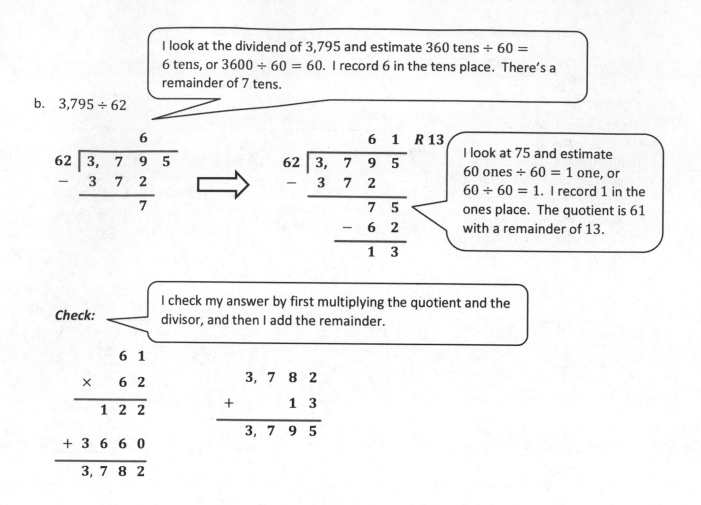

I look at 75 and estimate 60 ones ÷ 60 = 1 one, or 60 ÷ 60 = 1. I record 1 in the ones place. The quotient is 61 with a remainder of 13.

Check:

I check my answer by first multiplying the quotient and the divisor, and then I add the remainder.

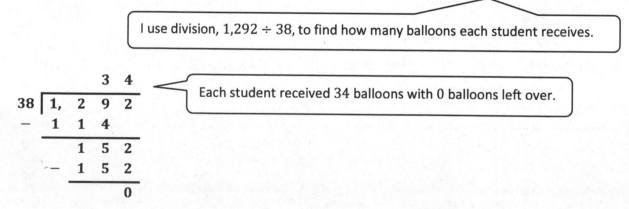

2.   1,292 balloons were shared equally among 38 students. How many balloons did each student receive?

I use division, 1,292 ÷ 38, to find how many balloons each student receives.

```
        3  4
38 | 1, 2  9  2
   -  1  1  4
         1  5  2
       -  1  5  2
                0
```

Each student received 34 balloons with 0 balloons left over.

Each student received 34 balloons.

Lesson 23:   Divide three- and four-digit dividends by two-digit divisors resulting in two- and three-digit quotients, reasoning about the decomposition of successive remainders in each place value.

© 2018 Great Minds®. eureka-math.org

Name _____ Date _____

1. Divide. Then, check using multiplication.

a. 9,962 ÷ 41

b. 1,495 ÷ 45

c. 6,691 ÷ 28

d. 2,625 ÷ 32

e. 2,409 ÷ 19

f. 5,821 ÷ 62

Lesson 23: Divide three- and four-digit dividends by two-digit divisors resulting in two- and three-digit quotients, reasoning about the decomposition of successive remainders in each place value.

© 2018 Great Minds®. eureka-math.org

183

2. A political gathering in South America was attended by 7,910 people. Each of South America's 14 countries was equally represented. How many representatives attended from each country?

3. A candy company packages caramel into containers that hold 32 fluid ounces. In the last batch, 1,848 fluid ounces of caramel were made. How many containers were needed for this batch?

Lesson 23: Divide three- and four-digit dividends by two-digit divisors resulting in two- and three-digit quotients, reasoning about the decomposition of successive remainders in each place value.

© 2018 Great Minds®. eureka-math.org

1. Divide.

   a. $3.5 \div 7 = \mathbf{0.5}$

   > I can use the basic fact of $35 \div 7 = 5$ to help me solve this problem. 3.5 is 35 tenths. 35 tenths $\div 7 = 5$ tenths, or 0.5.

   > Dividing by 70 is the same as dividing by 10 and then dividing by 7.

   b. $3.5 \div 70 = \mathbf{3.5 \div 10 \div 7}$
   $= \mathbf{0.35 \div 7}$
   $= \mathbf{0.05}$

   > 35 tenths $\div 10 = 35$ hundredths, or 0.35.

   > 35 hundredths $\div 7 = 5$ hundredths, or 0.05.

   c. $4.84 \div 2 = \mathbf{2.42}$

   > $4.84 = 4$ ones $+ 8$ tenths $+ 4$ hundredths.
   > 4 ones $\div 2 = 2$ ones, or 2.
   > 8 tenths $\div 2 = 4$ tenths, or 0.4.
   > 4 hundredths $\div 2 = 2$ hundredths, or 0.02.
   > The answer is $2 + 0.4 + 0.02 = 2.42$.

   > Dividing by 200 is equal to dividing by 100 and then dividing by 2.
   > Or I can think of it as dividing by 2 and then dividing by 100.

   d. $48.4 \div 200 = \mathbf{48.4 \div 2 \div 100}$
   $= \mathbf{24.2 \div 100}$
   $= \mathbf{0.242}$

   > $48 \div 2 = 24$
   > 4 tenths $\div 2 = 2$ tenths or 0.2.
   > So, $48.4 \div 2 = 24.2$.

   > I can visualize a place value chart. When I divide by 100, each digit shifts 2 places to the right.

Lesson 24: Divide decimal dividends by multiples of 10, reasoning about the placement of the decimal point and making connections to a written method.

© 2018 Great Minds®. eureka-math.org

185

2. Use place value reasoning and the first quotient to compute the second quotient. Use place value to explain how you placed the decimal point.

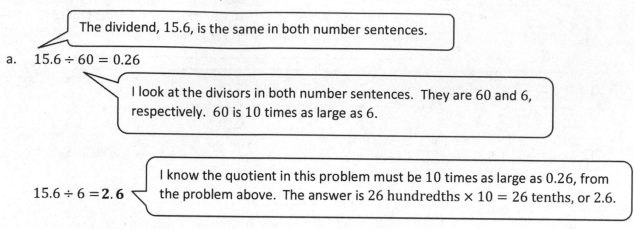

> The dividend, 15.6, is the same in both number sentences.

a. $15.6 \div 60 = 0.26$

> I look at the divisors in both number sentences. They are 60 and 6, respectively. 60 is 10 times as large as 6.

$15.6 \div 6 = 2.6$

> I know the quotient in this problem must be 10 times as large as 0.26, from the problem above. The answer is 26 hundredths × 10 = 26 tenths, or 2.6.

There are 10 *times fewer groups*, so there has to be 10 *times more in each group*.

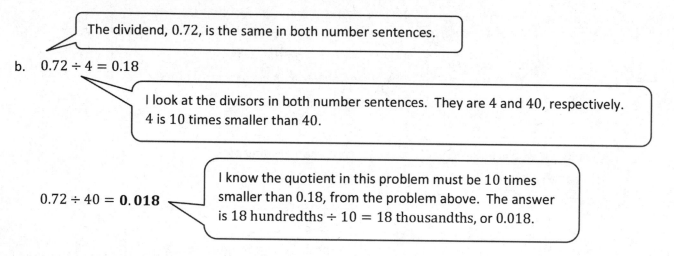

> The dividend, 0.72, is the same in both number sentences.

b. $0.72 \div 4 = 0.18$

> I look at the divisors in both number sentences. They are 4 and 40, respectively. 4 is 10 times smaller than 40.

$0.72 \div 40 = 0.018$

> I know the quotient in this problem must be 10 times smaller than 0.18, from the problem above. The answer is 18 hundredths ÷ 10 = 18 thousandths, or 0.018.

Instead of 4 groups, there are 40 groups. That's 10 *times more groups*, so there must be 10 *times less in each group*.

Lesson 24: Divide decimal dividends by multiples of 10, reasoning about the placement of the decimal point and making connections to a written method.

© 2018 Great Minds®. eureka-math.org

EUREKA MATH®

Name _____   Date _____

1. Divide. Show every other division sentence in two steps. The first two have been done for you.

   a. $1.8 \div 6 = 0.3$

   b. $1.8 \div 60 = (1.8 \div 6) \div 10 = 0.3 \div 10 = 0.03$

   c. $2.4 \div 8 =$ _____

   d. $2.4 \div 80 =$ _____

   e. $14.6 \div 2 =$ _____

   f. $14.6 \div 20 =$ _____

   g. $0.8 \div 4 =$ _____

   h. $80 \div 400 =$ _____

   i. $0.56 \div 7 =$ _____

   j. $0.56 \div 70 =$ _____

   k. $9.45 \div 9 =$ _____

   l. $9.45 \div 900 =$ _____

EUREKA MATH

Lesson 24:   Divide decimal dividends by multiples of 10, reasoning about the placement of the decimal point and making connections to a written method.

© 2018 Great Minds®. eureka-math.org

187

2. Use place value reasoning and the first quotient to compute the second quotient. Use place value to explain how you placed the decimal point.

a. $65.6 \div 80 = 0.82$

$65.6 \div 8 =$ _____

b. $2.5 \div 50 = 0.05$

$2.5 \div 5 =$ _____

c. $19.2 \div 40 = 0.48$

$19.2 \div 4 =$ _____

d. $39.6 \div 6 = 6.6$

$39.6 \div 60 =$ _____

Lesson 24: Divide decimal dividends by multiples of 10, reasoning about the placement of the decimal point and making connections to a written method.

© 2018 Great Minds®. eureka-math.org

3. Chris rode his bike along the same route every day for 60 days. He logged that he had gone exactly 127.8 miles.

    a.   How many miles did he bike each day? Show your work to explain how you know.

    b.   How many miles did he bike over the course of two weeks?

4. 2.1 liters of coffee were equally distributed to 30 cups. How many milliliters of coffee were in each cup?

Lesson 24:   Divide decimal dividends by multiples of 10, reasoning about the placement of the decimal point and making connections to a written method.

© 2018 Great Minds®. eureka-math.org

189

1.   Estimate the quotients.

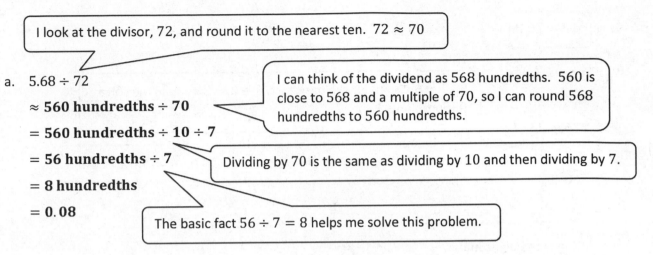

I look at the divisor, 72, and round it to the nearest ten. $72 \approx 70$

a. $5.68 \div 72$

I can think of the dividend as 568 hundredths. 560 is close to 568 and a multiple of 70, so I can round 568 hundredths to 560 hundredths.

$\approx 560 \text{ hundredths} \div 70$

$= 560 \text{ hundredths} \div 10 \div 7$

$= 56 \text{ hundredths} \div 7$

Dividing by 70 is the same as dividing by 10 and then dividing by 7.

$= 8 \text{ hundredths}$

$= 0.08$

The basic fact $56 \div 7 = 8$ helps me solve this problem.

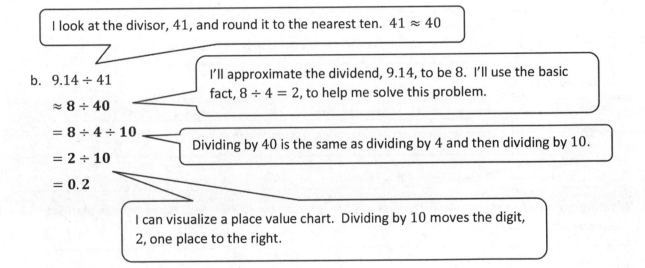

I look at the divisor, 41, and round it to the nearest ten. $41 \approx 40$

b. $9.14 \div 41$

I'll approximate the dividend, 9.14, to be 8. I'll use the basic fact, $8 \div 4 = 2$, to help me solve this problem.

$\approx 8 \div 40$

$= 8 \div 4 \div 10$

Dividing by 40 is the same as dividing by 4 and then dividing by 10.

$= 2 \div 10$

$= 0.2$

I can visualize a place value chart. Dividing by 10 moves the digit, 2, one place to the right.

Lesson 25:   Use basic facts to approximate decimal quotients with two-digit divisors,   **191**
                reasoning about the placement of the decimal point.

© 2018 Great Minds®. eureka-math.org

2. Estimate the quotient in (a). Use your estimated quotient to estimate (b) and (c).

$18 \approx 20$

a. $5.29 \div 18$

    $\approx 6 \div 20$

    $= 6 \div 2 \div 10$

    $= 3 \div 10$

    $= 0.3$

> $5.29 \approx 6$. I can use the basic fact, $6 \div 2 = 3$, to help me solve this problem.

> Dividing by 20 is the same as dividing by 2 and then dividing by 10.

> Since the digits in this expression are the same as (a), I can use place value understanding to help me solve.

b. $529 \div 18$

    $\approx 600 \div 20$

    $= 60 \div 2$

    $= 30$

> I can use the same basic fact, $6 \div 2 = 3$, to help me solve.

> $18 \approx 20$ and $529 \approx 600$

> $600 \div 20$ is equal to $60 \div 2$ because I divided both the dividend and the divisor by 10.

> My quotient makes sense! When I compare (b) to (a), I see that 529 is 100 times greater than 5.29. Therefore, the quotient should be 100 times greater as well. 30 is 100 times greater than 0.3.

c. $52.9 \div 18$

    $\approx 60 \div 20$

    $= 6 \div 2$

    $= 3$

> Again, I can use the same basic fact, $6 \div 2 = 3$, to help me solve this problem.

> I'll round 18 to 20 and approximate 52.9 to 60.

> $60 \div 20$ is equal to $6 \div 2$ because I divided both the dividend and the divisor by 10.

Lesson 25:   Use basic facts to approximate decimal quotients with two-digit divisors, reasoning about the placement of the decimal point.

© 2018 Great Minds®. eureka-math.org

Name _____   Date _____

1. Estimate the quotients.

   a.   3.53 ÷ 51 ≈

   b.   24.2 ÷ 42 ≈

   c.   9.13 ÷ 23 ≈

   d.   79.2 ÷ 39 ≈

   e.   7.19 ÷ 58 ≈

2. Estimate the quotient in (a). Use your estimated quotient to estimate (b) and (c).

   a.   9.13 ÷ 42 ≈

   b.   913 ÷ 42 ≈

   c.   91.3 ÷ 42 ≈

Lesson 25:   Use basic facts to approximate decimal quotients with two-digit divisors, reasoning about the placement of the decimal point.

© 2018 Great Minds®. eureka-math.org

193

3. Mrs. Huynh bought a bag of 3 dozen toy animals as party favors for her son's birthday party. The bag of toy animals cost $28.97. Estimate the price of each toy animal.

4. Carter drank 15.75 gallons of water in 4 weeks. He drank the same amount of water each day.

    a. Estimate how many gallons he drank in one day.

    b. Estimate how many gallons he drank in one week.

    c. About how many days altogether will it take him to drink 20 gallons?

Lesson 25:   Use basic facts to approximate decimal quotients with two-digit divisors,
reasoning about the placement of the decimal point.

© 2018 Great Minds®. eureka-math.org

1. Divide. Then check your work with multiplication.

    a. $48.07 \div 19 = \mathbf{2.53}$

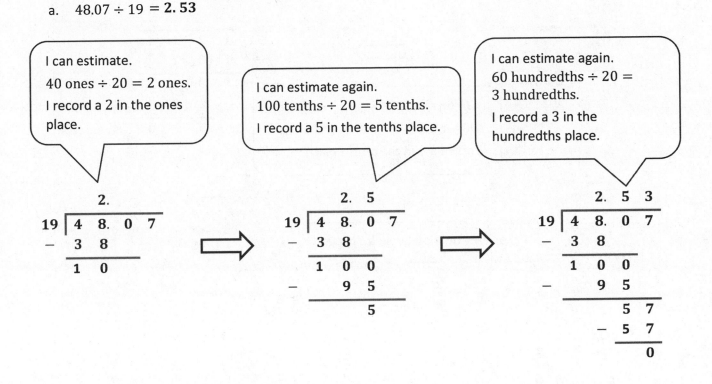

I can estimate.
40 ones ÷ 20 = 2 ones.
I record a 2 in the ones place.

I can estimate again.
100 tenths ÷ 20 = 5 tenths.
I record a 5 in the tenths place.

I can estimate again.
60 hundredths ÷ 20 = 3 hundredths.
I record a 3 in the hundredths place.

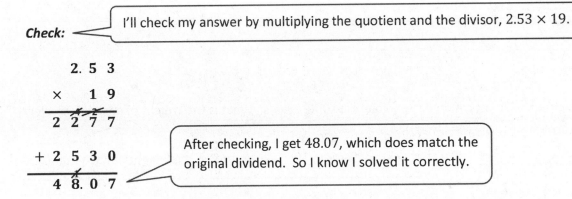

Check:

I'll check my answer by multiplying the quotient and the divisor, 2.53 × 19.

After checking, I get 48.07, which does match the original dividend. So I know I solved it correctly.

EUREKA MATH

Lesson 26: Divide decimal dividends by two-digit divisors, estimating quotients, reasoning about the placement of the decimal point, and making connections to a written method.

195

© 2018 Great Minds®. eureka-math.org

b. $122.4 \div 51$

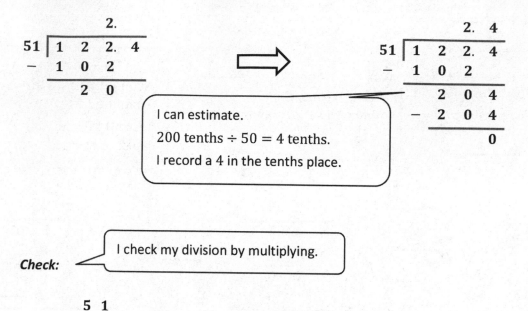

I can estimate.

200 tenths ÷ 50 = 4 tenths.

I record a 4 in the tenths place.

Check: I check my division by multiplying.

```
      5 1
  ×   2. 4
  ───────
    2 0 4
+ 1 0 2 0
  ───────
  1 2 2. 4
```

2. The weight of 42 identical mini toy soldiers is 109.2 grams. What is the weight of each toy soldier?

```
         2. 6
  42 | 1 0 9. 2
   −     8 4
     ───────
         2 5 2
     −   2 5 2
     ───────
             0
```

I can use division, $109.2 \div 42$, to find the weight of each toy soldier.

109.2 grams divided by 42 is equal to 2.6 grams with 0 grams remaining.

The weight of each toy soldier is 2. 6 grams.

Lesson 26: Divide decimal dividends by two-digit divisors, estimating quotients, reasoning about the placement of the decimal point, and making connections to a written method.

© 2018 Great Minds®. eureka-math.org

Name _____ Date _____

1. Create two whole number division problems that have a quotient of 9 and a remainder of 5. Justify which is greater using decimal division.

2. Divide. Then, check your work with multiplication.

   a. 75.9 ÷ 22

   b. 97.28 ÷ 19

   c. 77.14 ÷ 38

   d. 12.18 ÷ 29

Lesson 26: Divide decimal dividends by two-digit divisors, estimating quotients, reasoning about the placement of the decimal point, and making connections to a written method.

© 2018 Great Minds®. eureka-math.org

197

3. Divide.

     a. $97.58 \div 34$

                                                                         b. $55.35 \div 45$

4. Use the equations on the left to solve the problems on the right. Explain how you decided where to place the decimal in the quotient.

     a. $520.3 \div 43 = 12.1$ $52.03 \div 43 =$ _____

     b. $19.08 \div 36 = 0.53$ $190.8 \div 36 =$ _____

Lesson 26:   Divide decimal dividends by two-digit divisors, estimating quotients, reasoning about the placement of the decimal point, and making connections to a written method.

© 2018 Great Minds®. eureka-math.org

5. You can look up information on the world's tallest buildings at
   http://www.infoplease.com/ipa/A0001338.html.

   a. The Aon Centre in Chicago, Illinois, is one of the world's tallest buildings. Built in 1973, it is 1,136 feet
      high and has 80 stories. If each story is of equal height, how tall is each story?

   b. Burj al Arab Hotel, another one of the world's tallest buildings, was finished in 1999. Located in
      Dubai, it is 1,053 feet high with 60 stories. If each floor is the same height, how much taller or shorter
      is each floor than the height of the floors in the Aon Center?

EUREKA
MATH

Lesson 26:   Divide decimal dividends by two-digit divisors, estimating quotients,
reasoning about the placement of the decimal point, and making
connections to a written method.

© 2018 Great Minds®. eureka-math.org

199

1. Divide. Check your work with multiplication.

   $6.3 \div 18$

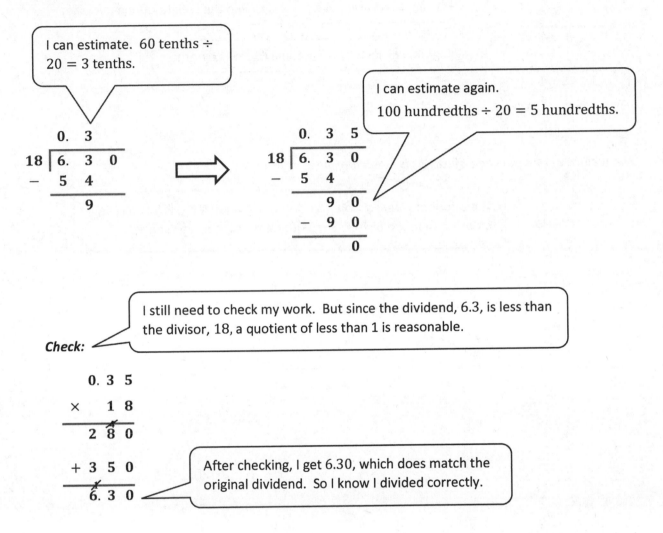

I can estimate. 60 tenths ÷ 20 = 3 tenths.

I can estimate again.

100 hundredths ÷ 20 = 5 hundredths.

```
   0. 3
18 6. 3  0
 -  5  4
        9
```

```
   0. 3  5
18 6. 3  0
 -  5  4
        9  0
     -  9  0
           0
```

I still need to check my work. But since the dividend, 6.3, is less than the divisor, 18, a quotient of less than 1 is reasonable.

Check:

```
    0. 3  5
  ×    1  8
  ────────
    2  8  0
  + 3  5  0
  ────────
    6. 3  0
```

After checking, I get 6.30, which does match the original dividend. So I know I divided correctly.

Lesson 27: Divide decimal dividends by two-digit divisors, estimating quotients, reasoning about the placement of the decimal point, and making connections to a written method.

201

EUREKA MATH

2. 43.4 kilograms of raisins was placed into 31 packages of equal weight. What is the weight of one package of raisins?

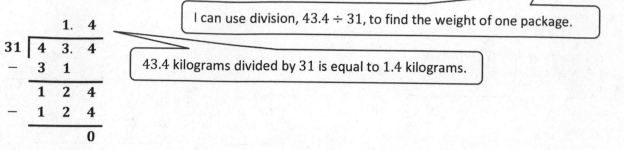

```
        1.  4
   31  4  3.  4
    −  3  1
       1  2  4
    −  1  2  4
             0
```

I can use division, 43.4 ÷ 31, to find the weight of one package.

43.4 kilograms divided by 31 is equal to 1.4 kilograms.

The weight of one package of raisins is 1. 4 kilograms.

The quotient is reasonable. Since the dividend, 43.4, is just a little bit more than the divisor, 31, a quotient of 1.4 makes sense.

Lesson 27: Divide decimal dividends by two-digit divisors, estimating quotients, reasoning about the placement of the decimal point, and making connections to a written method.

© 2018 Great Minds®. eureka-math.org

EUREKA MATH

Name _____   Date _____

1. Divide. Check your work with multiplication.

   a. 7 ÷ 28

   b. 51 ÷ 25

   c. 6.5 ÷ 13

   d. 132.16 ÷ 16

   e. 561.68 ÷ 28

   f. 604.8 ÷ 36

2. In a science class, students water a plant with the same amount of water each day for 28 consecutive days. If the students use a total of 23.8 liters of water over the 28 days, how many liters of water did they use each day? How many milliliters did they use each day?

Lesson 27: Divide decimal dividends by two-digit divisors, estimating quotients, reasoning about the placement of the decimal point, and making connections to a written method.

© 2018 Great Minds®. eureka-math.org

203

3. A seamstress has a piece of cloth that is 3 yards long. She cuts it into shorter lengths of 16 inches each. How many of the shorter pieces can she cut?

4. Jenny filled 12 pitchers with an equal amount of lemonade in each. The total amount of lemonade in the 12 pitchers was 41.4 liters. How many liters of lemonade would be in 7 pitchers?

Lesson 27: Divide decimal dividends by two-digit divisors, estimating quotients, reasoning about the placement of the decimal point, and making connections to a written method.
© 2018 Great Minds®. eureka-math.org

EUREKA MATH

1. Juanita is saving for a new television that costs $931. She has already saved half of the money. Juanita earns $19.00 per hour. How many hours must Juanita work to save the rest of the money?

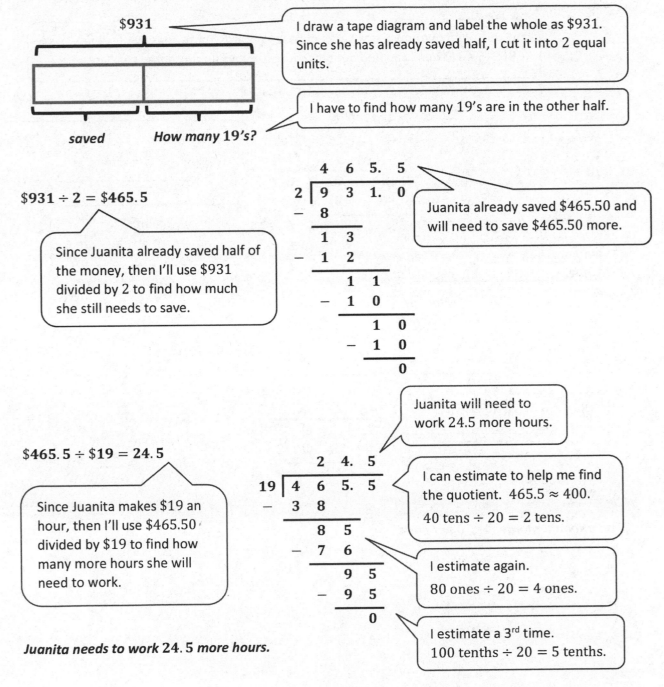

$931

I draw a tape diagram and label the whole as $931. Since she has already saved half, I cut it into 2 equal units.

saved   **How many 19's?**

I have to find how many 19's are in the other half.

$931 \div 2 = \$465.5$

Since Juanita already saved half of the money, then I'll use $931 divided by 2 to find how much she still needs to save.

Juanita already saved $465.50 and will need to save $465.50 more.

Juanita will need to work 24.5 more hours.

$465.5 \div \$19 = 24.5$

Since Juanita makes $19 an hour, then I'll use $465.50 divided by $19 to find how many more hours she will need to work.

I can estimate to help me find the quotient. $465.5 \approx 400$.

40 tens ÷ 20 = 2 tens.

I estimate again.
80 ones ÷ 20 = 4 ones.

I estimate a 3rd time.
100 tenths ÷ 20 = 5 tenths.

Juanita needs to work 24.5 more hours.

Lesson 28:   Solve division word problems involving multi-digit division with group size unknown and the number of groups unknown.

© 2018 Great Minds®. eureka-math.org

205

2. Timmy has a collection of 1,008 baseball cards. He hopes to sell the collection in packs of 48 cards and make $178.50 when all the packs are sold. If each pack is priced the same, how much should Timmy charge per pack?

> I need to find out how many packs of baseball cards Timmy has by dividing 1,008 ÷ 48. Then I can find out how much Timmy should charge per pack.

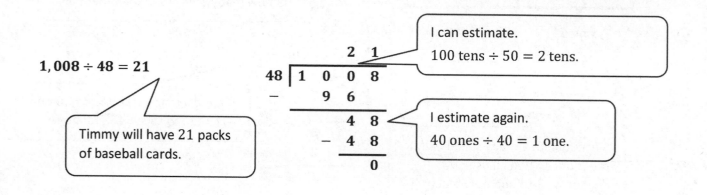

1,008 ÷ 48 = 21

> Timmy will have 21 packs of baseball cards.

> I can estimate.
> 100 tens ÷ 50 = 2 tens.

> I estimate again.
> 40 ones ÷ 40 = 1 one.

```
        2   1
48 | 1  0   0   8
  −    9    6
       ‾‾‾‾‾‾‾‾
            4   8
          − 4   8
            ‾‾‾‾‾
                0
```

$178.50 ÷ 21 = $8.50

> The price of each pack of cards needs to be $8.50.

```
            8.  5
21 | 1   7  8.  5
  −  1   6  8
     ‾‾‾‾‾‾‾‾
            1   0   5
          − 1   0   5
            ‾‾‾‾‾‾‾‾‾
                    0
```

Timmy should charge $8.50 per pack.

Lesson 28:   Solve division word problems involving multi-digit division with group size unknown and the number of groups unknown.

© 2018 Great Minds®. eureka-math.org

EUREKA
MATH

Name _____ Date _____

1. Mr. Rice needs to replace the 166.25 ft of edging on the flower beds in his backyard. The edging is sold in lengths of 19 ft each. How many lengths of edging will Mr. Rice need to purchase?

2. Olivia is making granola bars. She will use 17.9 ounces of pistachios, 12.6 ounces of almonds, 12.5 ounces of walnuts, and 12.5 ounces of cashews. This amount makes 25 bars. How many ounces of nuts are in each granola bar?

Lesson 28: Solve division word problems involving multi-digit division with group size unknown and the number of groups unknown.

207

© 2018 Great Minds®. eureka-math.org

3. Adam has 16.45 kg of flour, and he uses 6.4 kg to make hot cross buns. The remaining flour is exactly enough to make 15 batches of scones. How much flour, in kg, will be in each batch of scones?

4. There are 90 fifth-grade students going on a field trip. Each student gives the teacher $9.25 to cover admission to the theater and for lunch. Admission for all of the students will cost $315, and each student will get an equal amount to spend on lunch. How much will each fifth grader get to spend on lunch?

5. Ben is making math manipulatives to sell. He wants to make at least $450. Each manipulative costs $18 to make. He is selling them for $30 each. What is the minimum number he can sell to reach his goal?

Lesson 28:   Solve division word problems involving multi-digit division with group size unknown and the number of groups unknown.

© 2018 Great Minds®. eureka-math.org

1. Alonzo has 2,580.2 kilograms of apples to deliver in equal amounts to 19 stores. Eleven of the stores are in Philadelphia. How many kilograms of apples will be delivered to stores in Philadelphia?

$2,580.2 \div 19 = 135.8$

> I can use division to find out how many kilograms of apples are delivered to each store. Each store receives 135.8 kilograms of apples.

```
            1  3  5.  8
    19 | 2  5  8  0.  2
       -  1  9
             6  8
          -  5  7
             1  1  0
          -     9  5
                1  5  2
             -  1  5  2
                      0
```

$135.8 \times 11 = 1,493.8$

> Since I know each store receives 135.8 kilograms of apples, then I use multiplication to find the total kilograms of apples that will be delivered to 11 stores in Philadelphia.

```
         1  3  5. 8
      ×        1  1
         1  3  5  8
   +  1  3  5  8  0
      1  4  9  3. 8
```

1493.8 *kilograms of apples will be delivered to stores in Philadelphia.*

Lesson 29: Solve division word problems involving multi-digit division with group size unknown and the number of groups unknown.

209

© 2018 Great Minds®. eureka-math.org

2. The area of a rectangle is 88.4 m². If the length is 13 m, what is its perimeter?

> In order to find the perimeter, I need to know the width of the rectangle.

area = length × width

width = area ÷ length

$$= 88.4 \text{ m}^2 \div 13 \text{ m}$$

$$= 6.8 \text{ m}$$

```
        6.  8
13 | 8   8.  4
  −  7   8
        1 · 0   4
      −  1   0   4
                 0
```

> I know the width is equal to the area divided by the length. The width of the rectangle is 6.8 meters.

Perimeter of a rectangle = length + length + width + width

$$= 13 \text{ m} + 13 \text{ m} + 6.8 \text{ m} + 6.8 \text{ m}$$

$$= 26 \text{ m} + 13.6 \text{ m}$$

$$= 39.6 \text{ m}$$

```
    1 3.  0
    1 3.  0
       6.  8
  +    6.  8
  ─────────────
    3 9.  6
```

> I can add up all four sides of the rectangle to find the perimeter.

The perimeter of the rectangle is 39.6 meters.

Lesson 29: Solve division word problems involving multi-digit division with group size unknown and the number of groups unknown.

© 2018 Great Minds®. eureka-math.org

EUREKA MATH

Name _____   Date _____

Solve.

1. Michelle wants to save $150 for a trip to the Six Flags amusement park. If she saves $12 each week, how many weeks will it take her to save enough money for the trip?

2. Karen works for 85 hours throughout a two-week period. She earns $1,891.25 throughout this period. How much does Karen earn for 8 hours of work?

Lesson 29:   Solve division word problems involving multi-digit division with group
               size unknown and the number of groups unknown.

© 2018 Great Minds®. eureka-math.org

211

3. The area of a rectangle is 256.5 m². If the length is 18 m, what is the perimeter of the rectangle?

4. Tyler baked 702 cookies. He sold them in boxes of 18. After selling all of the boxes of cookies for the same amount each, he earned $136.50. What was the cost of one box of cookies?

5. A park is 4 times as long as it is wide. If the distance around the park is 12.5 kilometers, what is the area of the park?

Lesson 29:   Solve division word problems involving multi-digit division with group size unknown and the number of groups unknown.

© 2018 Great Minds®. eureka-math.org

EUREKA
MATH

Credits

Great Minds® has made every effort to obtain permission for the reprinting of all copyrighted material. If any owner of copyrighted material is not acknowledged herein, please contact Great Minds for proper acknowledgment in all future editions and reprints of this module.